Finite Element Modeling of Multiscale Transport Phenomena

Finite Element Modeling of Multiscale Transport Phenomena

Vahid Nassehi
Loughborough University, UK

Mahmoud Parvazinia
Iran Polymer and Petrochemical Institute, Iran

ICP

Imperial College Press

Published by

Imperial College Press
57 Shelton Street
Covent Garden
London WC2H 9HE

Distributed by

World Scientific Publishing Co. Pte. Ltd.
5 Toh Tuck Link, Singapore 596224
USA office: 27 Warren Street, Suite 401-402, Hackensack, NJ 07601
UK office: 57 Shelton Street, Covent Garden, London WC2H 9HE

British Library Cataloguing-in-Publication Data
A catalogue record for this book is available from the British Library.

FINITE ELEMENT MODELING OF MULTISCALE PHENOMENA

Desk Editor: Tjan Kwang Wei

ISBN-13 978-1-84816-429-1
ISBN-10 1-84816-429-7

Typeset by Stallion Press
Email: enquiries@stallionpress.com

Printed in Singapore.

For our families

Preface

For more than five decades finite element method has been regarded as the foremost numerical technique for the solution of governing equations of engineering problems. Using this method powerful computer schemes, which combine mathematical rigor with geometrical flexibility in dealing with complex problems, have been developed. These schemes are, in general, the most reliable predictive simulation tools available to researchers and design engineers. Inherent flexibility and sound mathematical basis of the technique has been the main reason for its progress and evolution over the years, enabling numerical analysts to use the technique to solve a wide variety of problems. Recent advances in the power and capability of computer systems has made simulation of some of the most complex physical situations a realistic possibility. However, novel concepts need to be incorporated into existing numerical schemes in order to take advantage of recent advances in low cost computing. In this respect, the solution of multiscale phenomena, in which significant and rapid variations in the behavior of field unknowns either precludes the application of traditional techniques or renders their use complicated and unyielding, has been the subject of intense research during the past decade. During this period many important research papers have appeared which provide robust mathematical foundations for the construction of practical finite element schemes for multiscale problems.

The main focus of this book is to provide a simple to follow account of the development of a class of practical multiscale weighted residual finite element schemes for field problems encountered in fluid flow and transport processes. In particular, dealing with the generic multiscale phenomena which affect the design and analysis of chemical engineering and polymer processing operations has been our objective.

The book starts with an explanation of the weighted residuals finite element technique to provide the necessary background for the discussions presented later on in the book. Readers who have not previously used weighted residuals finite element schemes should, nevertheless, be able to follow the discussions presented in chapters dealing with the extension of this technique to multiscale problems. Almost all of the topics introduced in the text have been supplemented with solved examples. These examples can be used as a guide by readers to apply the constructed schemes to their own problems or they may use the described methodology for the development of multiscale schemes applicable to other problems.

Finally, we have included a detailed listing of the computer code used to solve many of the examples given in this book and provided sample input and output files. Readers can repeat the illustrated examples and gain experience for extending the program to perform their own multiscale finite element simulations.

Vahid Nassehi
Mahmoud Parvazinia

Contents

CHAPTER 1

Weighted Residual Finite Element Method

There are a number of ways in which a finite element solution scheme for engineering problems can be constructed. Some of these techniques are problem specific and do not have universal applicability. For example, the displacement method commonly used by civil and mechanical engineers is more appropriate to cases where the problem requiring a solution can be stated in terms of a variational principle. Most types of process engineering problems, however, can only be formulated in terms of a set of governing differential equations derived from fundamental laws of physics. Therefore the development of finite element schemes in process engineering requires a more general approach. Weighted residual technique provides such a general approach for the development of efficient finite element schemes for the solution of the governing partial differential equations that represent many types of process engineering problems.

In this chapter, the basic concepts underpinning this approach are explained. We will mainly focus on the methodology and procedures used for the development of numerical solution schemes and will avoid discussions related to the fundamental mathematical theory of the method.

1.1 Basic Concept

Numerical solution of differential equations arising in engineering problems is usually based on finite difference, finite element, boundary element or finite volume techniques. Other numerical methods may also be used to solve specific problems. In general, the finite element method has a greater geometrical flexibility than other currently available numerical methods. It can also cope very effectively with a wide range of boundary conditions (Nassehi, 2002). The general weighted residual method is the basic technique used to construct finite element schemes for field problems. Therefore we start with a brief description of this method.

Consider a boundary value problem represented as

$$\begin{cases} LT = f & \text{in } \Omega \\ T = a & \text{on } \Gamma \end{cases} \tag{1.1}$$

where Ω is the problem domain and Γ is its boundary, in the absence of an analytical solution an approximate representation of T can be written as

$$T \approx \tilde{T} = a + \sum_{i=1}^{m} \alpha_i \varphi_i \tag{1.2}$$

where α_i is a set of constant coefficients and φ_i represents a set of geometrical functions called basis functions. Substituting Eq. (1.2) into Eq. (1.1) we have

$$L\tilde{T} - f = R_\Omega \tag{1.3}$$

where $R_\Omega \neq 0$ is the residual which will inevitably appear through the insertion of an approximation instead of an exact solution for the field variable into the differential equation. This equation can now be written as

$$L\left[a + \sum_{i=1}^{m} \alpha_i \varphi_i\right] - f = R_\Omega \tag{1.4}$$

The weighted residual method is based on the elimination of this residual. To achieve this, the residual is weighted by appropriate position-dependent functions and integrated over the domain to obtain a statement as

$$\int_\Omega W_j R_\Omega d\Omega = 0 \quad j = 1, 2, 3, \ldots, m \tag{1.5}$$

where W_j are linearly independent weight functions. The above equation can be rewritten as

$$\int_\Omega W_j \left\{ L\left[a + \sum_{i=1}^{m} \alpha_i \varphi_i\right] - f \right\} d\Omega = 0 \quad j = 1, 2, 3, \ldots, m \tag{1.6}$$

This equation represents the weighted residual statement of the original differential equation.

1.1.1 *Practical procedure*

As we see from the outlined concept, the weighted residual solution of the described problem depends on the use of appropriate weight and basis functions. In order to develop practical solution schemes for field problems these functions need to be determined in a systematic and rigoros manner.

The problem of finding weight functions is more straightforward and we explain it first. To simplify our explanations of the technique, here we use an example to outline the steps required to obtain a weighted residual statement for a typical partial differential equation.

Consider the following transient convection–diffusion equation

$$\frac{\partial f}{\partial t} + \mathbf{v} \cdot \nabla f - \nabla \cdot k \nabla f - \dot{q} = 0 \tag{1.7}$$

where f is the unknown function (i.e. the field variable), t is time, $\mathbf{v}$ is the velocity vector, k is a diffusion coefficient, and $\dot{q}$ is a source/sink term. We can start by approximating the time dependent term in Eq. (1.7) using a simple finite difference relationship such as a forward difference. Therefore

$$\frac{f^{n+1} - f^n}{\Delta t} + \mathbf{v} \cdot \nabla f^{n+\theta} - \nabla \cdot k \nabla f^{n+\theta} - \dot{q} = 0 \tag{1.8}$$

This means that the field variable in Eq. (1.8), given at a time level $n + \theta$, $0 \leq \theta \leq 1$, is discretized using a forward difference formula. If we now substitute the function f in Eq. (1.7) with an approximate form analogous to Eq. (1.2) an error is generated. Therefore representing $\tilde{f}$ in terms of basis functions (i.e. $\tilde{f} = \sum_{i=1}^{m} \varphi_i f_i$) the following least squares functional can be constructed

$$I_{n+1} = \int_\Omega \left[\frac{\sum_{i=1}^{m} \varphi_i f_i^{n+1} - \sum_{i=1}^{m} \varphi_i f_i^n}{\Delta t_n} + \mathbf{v} \cdot \nabla \left(\sum_{i=1}^{m} \varphi_i f_i^{n+\theta} \right) \right.$$

$$\left. - \nabla \cdot k \nabla \left(\sum_{i=1}^{m} \varphi_i f_i^{n+\theta} \right) - \dot{q} \right]^2 d\Omega = \int_\Omega [f_{\text{res}}^l]^2 d\Omega \tag{1.9}$$

where n is the time level. Minimization of functional (1.9) with respect to f_i gives

$$\frac{\partial I}{\partial f_i} = 2 \int_{\Omega_e} \left(\frac{\varphi_i}{\Delta t_n} + \theta \mathbf{v} \cdot \nabla \varphi_i - \theta \nabla \cdot k \nabla \varphi_i \right) [f_{\text{res}}^l] d\Omega = 0 \tag{1.10}$$

and

$$\int_\Omega (\varphi_i + \theta \Delta t_n \mathbf{v} \cdot \nabla \varphi_i - \theta \Delta t_n \nabla \cdot k \nabla \varphi_j)[f_{\text{res}}^l] d\Omega = 0 \tag{1.11}$$

or

$$\int_\Omega W[f_{\text{res}}^l] d\Omega = 0 \tag{1.12}$$

which is similar to Eq. (1.5). We see that Eq. (1.11) represents a weighted residual statement where the weighting function is given as

$$W = \varphi_i + \theta \Delta t_n \mathbf{v} \cdot \nabla \varphi_i - \theta \Delta t_n \nabla \cdot k \nabla \varphi_i \tag{1.13}$$

By selecting various terms of the weight function given in Eq. (1.13) the following weighted residual methods can be developed

- The standard (Babnov) Galerkin method is obtained if we only keep the first term in Eq. (1.13). It can hence be readily ascertained that in this method the basis and weight functions are identical.
- A first order Petrov–Galerkin scheme is obtained by neglecting the third term in Eq. (1.13). This method corresponds to a technique called Stream Line Upwind Petrov–Galerkin (SUPG) scheme, a special form of this scheme (called Inconsistent Upwinding) in which the second term in the weight function is only applied to the weighting of the convection term is also used (however, note that the inconsistent upwinding cannot be regarded as a true weighted residual method).
- If we retain all of the terms in the weight function, a scheme corresponding to a second order Petrov–Galerkin formulation is obtained.

In steady state problems the time-dependent term of the residual is eliminated. Therefore, for steady state problems a scheme equivalent to the combination of Galerkin and least square methods is obtained.

The procedure leading to the development of practical weighted residual schemes does not provide a systematic technique for the derivation or selection of appropriate basis functions. This has been a fundamental difficulty with the use and application of these theoretically powerful techniques in practical problems. However, combination of these concepts with the finite element discretization resolves all such problems in a very efficient manner. In the following sections we describe the procedures used to develop finite element approximations and give examples of the solution of differential equations by the weighted residuals finite element schemes.

1.1.2 *Finite element approximations*

The first step in the formulation of a finite element approximation for a field problem is to divide the problem domain into a number of smaller subregions without leaving any gaps or allowing any overlapping between them. This process is called "Domain Discretization." An individual sub-region in a discretized domain is called a "finite element" and collectively, the finite elements provide a "finite element mesh" for the discretized domain.

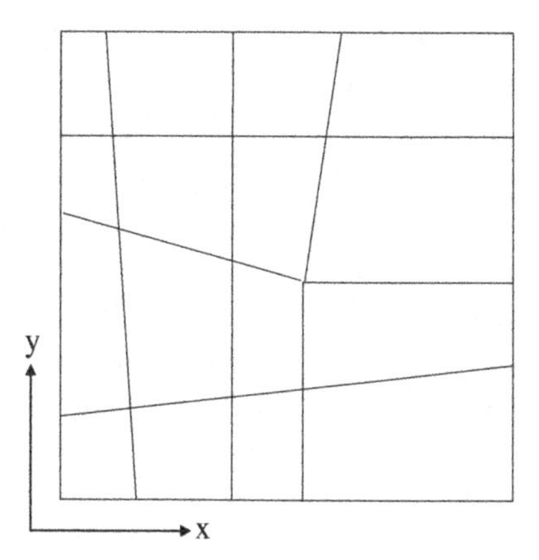

Fig. 1.1 A finite element mesh of four node quadrilateral elements.

In general, the elements in a finite element mesh may have different sizes but all of them usually have a common basic shape (e.g. they are all lines, triangles or quadrilaterals, prisms or tetrahedrons in one-, two-, and three-dimensional domains, respectively) and have an equal number of nodes (Fig. 1.1).

The nodes are the sampling points in an element where the numerical values of the unknowns will be calculated. All types of finite elements should have some nodes located on their boundary lines. Some of the commonly used finite elements also have interior nodes. Boundary nodes of the individual finite elements appear as the junction points between the elements in a finite element mesh.

In most engineering problems the boundary of the problem domain includes curved sections. The discretization of domains with curved boundaries using meshes that consist of elements with straight sides inevitably involves some error. This type of discretization error can obviously be reduced by mesh refinements. Mesh refinement simply means that the size of the elements that make up the mesh is reduced and a curved boundary can be approximated more accurately. However, in general, such a discretization error cannot be entirely eliminated unless finite elements which themselves have curved sides are used.

After the discretization of a problem domain into finite elements, an unknown field function, anywhere inside an element, can be approximately calculated as an interpolant of its values at the nodes of that element. Nodal values of a field unknown (called the nodal degrees of freedom) therefore appear as coefficients multiplied to geometrical interpolation (shape) functions. Obviously the form of geometrical interpolation functions associated with a finite element will depend on its shape and the number of nodes that can be selected as the sampling for that element. Different types of finite elements have therefore been developed over the past decades to generate desired approximations for a wide range of problems.

The described interpolation provides a systematic way of deriving approximate forms within a finite element for field unknowns in terms of their nodal values and set of geometrical functions. As these approximations are derived on an elemental space their use to form weighted residual statements results on an elemental statement. After this initial step, however, weighted residual statements for all of the elements in a computational grid can be formed and assembled together to obtain a global system of equations whose solution yields the values of nodal degrees of freedom.

Further explanations regarding the procedure for the derivation of interpolation functions, inter-element continuity of variables and elemental residual statements are explained in Chap. 2.

1.2 Numerical Integration

As demonstrated in Eq. (1.6), weighted residual solution of differential equations requires function integrations. However, after the domain discretization into finite elements the required integrations are confined to the space of individual elements. Therefore the functions that need to be integrated originate from geometrical interpolation functions. This means that the integrands are often low degree polynomials. Therefore they can be accurately evaluated using simple numerical integration (quadrature) techniques. The main advantage of using quadrature is that these techniques can be incorporated into a computer program rendering the entire finite element solution procedure automatic.

Consider the integration of a function $f(x_1, x_2)$ over a quadrilateral element within a finite element mesh, as shown in Fig. 1.2, we have

$$I = \int_{\alpha_1}^{\beta_1} \int_{\alpha_2}^{\beta_2} f(x_1, x_2) dx_1 dx_2 \tag{1.14}$$

where x_1 and x_2 are the global coordinates and α_1, α_2, β_1, and β_2 are the limits of integration.

Integration limits in Eq. (1.14) will change from element to element in a mesh if global coordinates are used. This makes the finite element procedure cumbersome. However, as shown in Fig. 1.2, mapping of elements from the global mesh onto a master element of regular shape (in this case

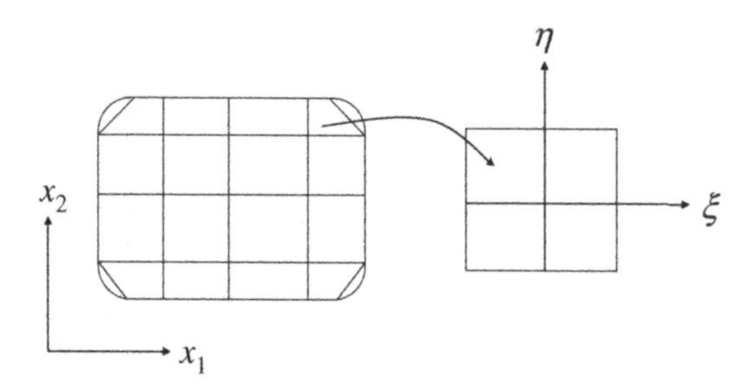

Fig. 1.2 Transformation from a global system to a master element with locally defined coordinates.

a square because the elements in the original mesh are two-dimensional quadrilaterals and can be mapped onto a square) resolves this difficulty. After such a transformation, the required integrals can first be calculated using uniform limits over the master element and then transformed back to the original system. Further simplifications can also be considered. For example, the use of a normalized local coordinate system corresponds to the uniform integration limits of -1 to $+1$. This allows the utilization of a simple numerical integration method called the Gauss–Legendre formula (Gerald and Wheatley, 1984).

An additional advantage of the described mapping is to overcome the problem of curved boundaries. Using elements which have straight sides inevitably results in discretization errors (see the gap between the corner elements in Fig. 1.2 and the actual domain boundary). However, elements with curved sides can also be mapped into a master element of regular shape. A fuller explanation of this point is outlined below.

1.2.1 *Mapping of irregular and curved elements onto master elements*

A one-to-one transformation (mapping) between any global and local coordinate systems, respectively representing actual finite element mesh and a selected master element, can be established using a variety of techniques. The most general method is a form of "parametric mapping" in which the transformation functions are the polynomials based on the element shape functions (Nassehi, 2002).

A graphical example of mapping between an irregular shape element and a regular master element is shown in Fig. 1.3. The transformation is

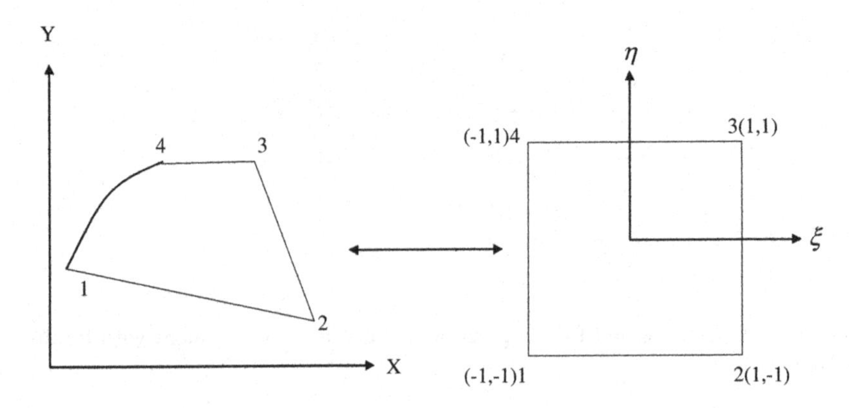

Fig. 1.3 Isoparametric mapping of an irregular element.

represented as

$$\mathrm{M} = \begin{cases} x = x(\xi, \eta) \\ y = y(\xi, \eta) \end{cases} \tag{1.15}$$

In "Isoparametric" mapping, which is the most commonly used form of parametric mapping, the coordinate transformation functions are identical to the interpolation functions associated with the selected finite elements. However, it should be noted that the linear shape functions in ξ and η can only be used to map irregular elements with straight sides to a master element.

In general, elements with curved sides can only be generated using higher-order master elements. Isoparametric transformation functions between a global coordinate system and local coordinates are, in general, written as

$$\begin{cases} x = \sum_{i=1}^{p} N_i(\xi, \eta) x_i \\ y = \sum_{i=1}^{p} N_i(\xi, \eta) y_i \end{cases} \tag{1.16}$$

where x_i and y_i are the nodal coordinates in the global system. Implementation of isoparametric mapping is based on the expression of the derivatives of shape functions in terms of local variables. Derivatives of a function T in terms of local variables (ξ, η) can be expressed in terms of its global derivatives, given with respect to global (x, y) coordinates, using the rules of differentiation as

$$\begin{cases} \dfrac{\partial T}{\partial \xi} = \dfrac{\partial T}{\partial x}\dfrac{\partial x}{\partial \xi} + \dfrac{\partial T}{\partial y}\dfrac{\partial y}{\partial \xi} \\ \dfrac{\partial T}{\partial \eta} = \dfrac{\partial T}{\partial x}\dfrac{\partial x}{\partial \eta} + \dfrac{\partial T}{\partial y}\dfrac{\partial y}{\partial \eta} \end{cases} \tag{1.17}$$

Therefore using matrix notation

$$\begin{bmatrix} \frac{\partial T}{\partial \xi} \\ \frac{\partial T}{\partial \eta} \end{bmatrix} = \begin{bmatrix} \frac{\partial x}{\partial \xi} & \frac{\partial y}{\partial \xi} \\ \frac{\partial x}{\partial \eta} & \frac{\partial y}{\partial \eta} \end{bmatrix} \begin{bmatrix} \frac{\partial T}{\partial x} \\ \frac{\partial T}{\partial y} \end{bmatrix} \tag{1.18}$$

or

$$\begin{bmatrix} \frac{\partial T}{\partial \xi} \\ \frac{\partial T}{\partial \eta} \end{bmatrix} = J \begin{bmatrix} \frac{\partial T}{\partial x} \\ \frac{\partial T}{\partial y} \end{bmatrix} \tag{1.19}$$

where matrix J is called the "Jacobian" of coordinate transformation. Therefore

$$\begin{bmatrix} \frac{\partial T}{\partial x} \\ \frac{\partial T}{\partial y} \end{bmatrix} = J^{-1} \begin{bmatrix} \frac{\partial T}{\partial \xi} \\ \frac{\partial T}{\partial \eta} \end{bmatrix} \tag{1.20}$$

we can hence write

$$\begin{bmatrix} \frac{\partial \tilde{T}}{\partial x} \\ \frac{\partial \tilde{T}}{\partial y} \end{bmatrix} = J^{-1} \begin{bmatrix} \dfrac{\partial \sum_{i=1}^{p} N_i(\xi,\eta)T_i}{\partial \xi} \\ \dfrac{\partial \sum_{i=1}^{p} N_i(\xi,\eta)T_i}{\partial \eta} \end{bmatrix} \tag{1.21}$$

Obviously the existence of an inverse for the Jacobian matrix is a necessary requirement for the described transformation. It should also be noted that the mapping of irregular elements into master elements generates some degree of approximation in finite element calculations. In general, mapping of highly distorted elements generates a high degree of mapping error and in extreme cases the sign of the Jacobian may change during transformation, rendering the operation invalid. In certain types of finite element approximations, transformation of second-order derivatives is necessary. Isoparametric transformation of second- or higher-order derivatives is not trivial and needs lengthy manipulations. Details of a procedure for transformation of second-order derivatives are represented in Petera *et al.* (1993).

Transformation of Eq. (1.14) into the square element shown in Fig. 1.3 gives

$$I = \int_{-1}^{1}\int_{-1}^{1} F(\xi,\eta)\det \mathbf{J}^e d\xi d\eta \tag{1.22}$$

where $\mathbf{J}^e$ is the Jacobian of coordinate transformation (Spiegel, 1974) and the limits of integration are defined by the local coordinates. Therefore

$$dx_1 dx_2 = \det \mathbf{J}^e d\xi d\eta \tag{1.23}$$

After some algebraic manipulations we can write

$$I = \int_{-1}^{1}\int_{-1}^{1} G(\xi,\eta)\, d\xi d\eta \tag{1.24}$$

According to the Gauss–Legendre quadrature I is calculated as

$$I = \int_{-1}^{1}\int_{-1}^{1} G(\xi,\eta)\, d\xi d\eta \approx \sum_{i=1}^{m}\sum_{j=1}^{n} G(\xi_i,\eta_j) w_i w_j \tag{1.25}$$

where ξ_i and η_j are the quadrature point coordinates, w_i and w_j are the corresponding weight factors and m and n are the number of quadrature points in each summation.

The number of quadrature points in these summations depends on the order of the polynomial function in the integral. In one-dimensional problems this quadrature yields an exact result for a polynomial of degree $2n-1$ (or less) using n points. Integrands in elemental equations are based on interpolation functions, which are low-order polynomials. Therefore the number of required quadrature points in practical calculations is low (usually $n = 2$ or 3).

The described example can be readily implemented in quadrilateral elements and its generalization to three-dimensional brick type elements with 8, 20 or 27 nodes is straightforward. However, there are many applications in which the domain discretization into quadrilateral elements or their three-dimensional counterparts creates difficulties. In these cases triangular and tetrahedral elements for two- and three-dimensional problems are used. It is not possible to choose an element-based coordinate system for triangular elements in which limits of integration remain between -1 and $+1$. Therefore, the described Gauss–Legendre quadrature is not a suitable method for such elements. However, this problem can be resolved using properly defined "area coordinates" in appropriately selected triangular master elements. Detailed definition of area coordinates and the technique for developing quadrature techniques for triangular elements are given by

Zienkiewicz and Taylor (1994) who list the coordinates of the sampling points and their corresponding weights. The coordinates and integration limits in these elements are therefore defined with respect to area coordinates. After the definition of the local coordinate system in terms of area coordinates a straightforward quadrature procedure for these elements can be constructed. Tables 1.1–1.3 show the sampling points and weighting factors, respectively, for quadrilateral, triangular, and tetrahedral elements.

Using the data provided in Table 1.2, a polynomial function of order $q(\xi^i\eta^i$ with $i+j \leq q)$ can be calculated as

$$\int_0^1 \int_0^{1-\xi} G(\xi,\eta)d\eta d\xi \cong \sum_{i=1}^{n} w_i G(\xi_i,\eta_i)$$

in which n in the number of quadrature points (Dhatt and Touzot, 1985).

Table 1.1. Quadrature points and weights for quadrilateral elements.

$m = n$	ξ_i, η_i	w_i
2	±0.57735027	1
3	0	0.88888889
	±0.77459667	0.55555555
4	±0.33998104	0.65214515
	±0.86113631	0.34785485
5	0	0.56888889
	±0.53846931	0.47862867
	±0.90617985	0.23692689
6	±0.23861918	0.46791393
	±0.66120939	0.36076157
	±0.93246951	0.17132449
7	0	0.41795918
	±0.40584515	0.38183005
	±0.74153119	0.27970539
	±0.94910791	0.12948497
8	±0.18343464	0.36268378
	±0.52553241	0.31370665
	±0.79666648	0.22238103
	±0.96028986	0.10122854
10	±0.14887434	0.29552422
	±0.43339539	0.26926672
	±0.67940957	0.21908636
	±0.86506337	0.14945135
	±0.97390653	0.06667134

Table 1.2. Quadrature points and weights for triangular elements.

Order m	Number of points	Coordinates ξ_i	Coordinates ηi	Weights, w_i
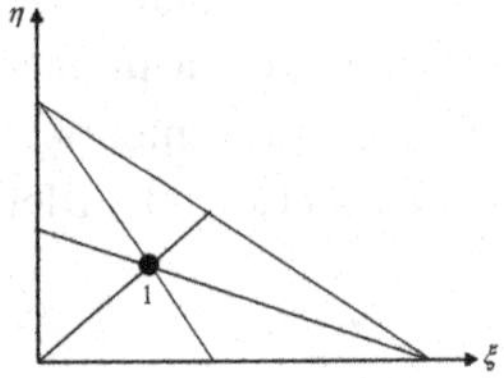 1	1	1/3	1/3	1/2
		1/2	1/2	
2	3	0	1/2	1/6
		1/2	0	
		1/6	1/6	
2	3	2/3	1/6	1/6
		1/6	2/3	
3	4	1/3	1/3	−27/96
		1/5	1/5	
		3/5	1/5	25/96
		1/5	3/5	
		a	a	
4	6	1–2a	a	0.11169079
		a	1–2a	
$a = 0.445948$				
		b	b	
$b = 0.091576$		1–2b	b	
		b	1–2b	0.05497587

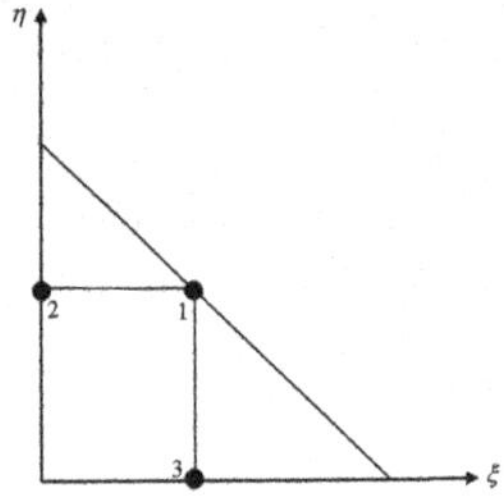

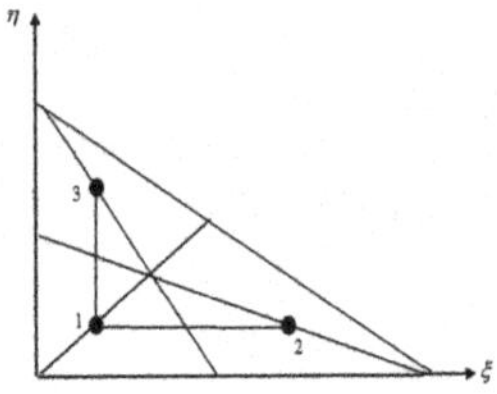

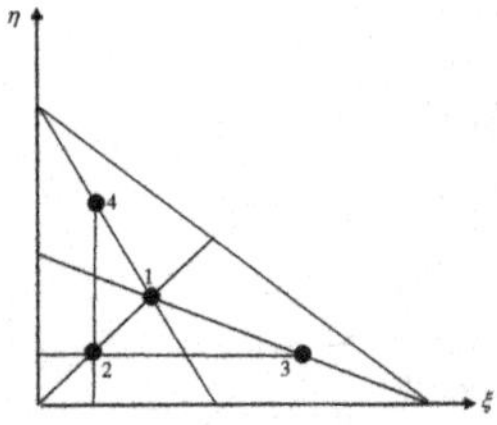

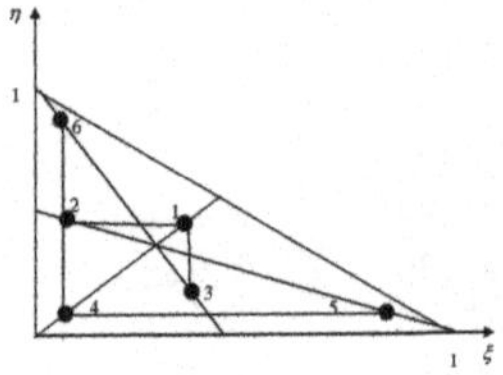

(Continued)

Table 1.2. (*Continued*)

Order m	Number of points	Coordinates ξ_i	Coordinates ηi	Weights, w_i
5	7	1/3	1/3	9/80
		a	a	
$a = \dfrac{6+\sqrt{15}}{21}$		1–2a	a	0.06619707
		a	1–2a	
$= 0.470142$				
		b	b	
$b = \dfrac{4}{7} - a$		1–2b	b	0.06296959
		b	1–2b	
$= 0.101285$				
		a	a	
		1–2a	a	0.02542245
6	12	a	1–2a	
$a = 0.063089$		b	b	
		1–2b	b	0.05839314
$b = 0.249286$		b	1–2b	
$c = 0.310352$		c	d	
		d	c	
$d = 0.053145$		$1-(c+d)$	c	0.04142554
		$1-(c+d)$	d	
		c	$1-(c+d)$	
		d	$1-(c+d)$	

The corresponding quadrature formula for a tetrahedral element is given as

$$\int_0^1 \int_0^{1-\xi} \int_0^{1-\xi-\eta} G(\xi, n, \zeta) d\zeta d\eta d\xi \cong \sum_{i=1}^{n} w_i G(\xi_i, \eta_i, \xi_i)$$

In which n is the number of quadrature points (Dhatt and Touzot, 1985).

1.3 Steps Used to Obtain a Finite Element Solution for a Field Problem

Based on the explanations we have given up to this point it is now possible to outline the steps used to develop a weighted residual finite element solution for a field problem.

Table 1.3. Quadrature points and weights for tetrahedral elements.

Order m	Number of points r	Coordinates ξ_i η_i ς_i	Weights w_i
1	1	$\frac{1}{4}$ $\frac{1}{4}$ $\frac{1}{4}$ $\frac{1}{4}$	$\frac{1}{6}$
2	4 $a = \frac{5-\sqrt{5}}{20}$ $b = \frac{5+3\sqrt{5}}{20}$	a a a a a b a b a b a a	$\frac{1}{24}$
3	5 $a = \frac{1}{4}$ $b = \frac{1}{6}$ $c = \frac{1}{2}$	a a a b b b b b c b c b c b b	$-\frac{2}{15}$ $\frac{3}{40}$
5	15 $a = \frac{1}{4}$ $a = \frac{1}{4}$ $\begin{matrix} b1 \\ b2 \end{matrix} = \frac{7-\sqrt{15}}{34}$ $\begin{matrix} c1 \\ c2 \end{matrix} = \frac{13+3\sqrt{15}}{34}$ $d = \frac{5-\sqrt{15}}{20}$ $e = \frac{5+\sqrt{15}}{20}$	a a a b_i b_i b_i b_i b_i b_i b_i b_i c_i b_i c_i b_i c_i b_i b_i d d e d e d e d d d e e e d e e e d	$\frac{8}{405}$ or $\frac{112}{5670}$ $i = 1, 2$ $\frac{w_1}{w_2} = \frac{2665 + 14\sqrt{15}}{226800}$ $\frac{5}{567}$

Step 1. The global solution domain is discretized into a number of suitable finite elements.
Step 2. The field variables in the governing equations are substituted by approximations in terms of the interpolation functions associated with the type of finite elements used to discretize the domain.
Step 3. An elemental weighted residual statement is constructed.
Step 4. If necessary, by the application of procedures such as integration by parts (or in general Green's theorem), the order of the

differentials in the formed weighted residual equations are reduced to make the approximations compatible with the elements used.

Step 5. Using matrix notations, elemental stiffness equations representing the described weighted residual forms are derived.

Step 6. Elemental stiffness equations are assembled into a global system of algebraic equations over their common nodes. In this step isoparametric mapping can be used to transform the derived equations into a local natural coordinate system. Note that this will make it possible to use quadrature to evaluate integrals in the elemental equations. Therefore, the procedure will be commonly employed irrespective of the shape of the elements in the global mesh.

Step 7. Boundary conditions are incorporated into the assembled global system of equations to make it determinate.

Step 8. The global set of equations is solved using an appropriate solution technique for the set of algebraic equations.

In the following sections we present a number of solved examples in which step-by-step application of the outlined procedure is described.

1.3.1 *Example 1*

We start with the standard Galerkin finite element solution of the following one-dimensional steady state problem. Consider a one-dimensional domain as shown in Fig. 1.4.

The following differential equation represents a molecular process (e.g. heat conduction) along this domain and its finite element solution should provide discrete numerical results for the field unknown (T) on a number of selected internal points.

$$\frac{d^2T}{dx^2} + T = 10 \quad \text{in} \quad \Omega \quad \text{Subject to} \tag{1.26}$$

$$T_A(x = 0) = 1 \quad \text{and} \quad T_B(x = 1) = 4$$

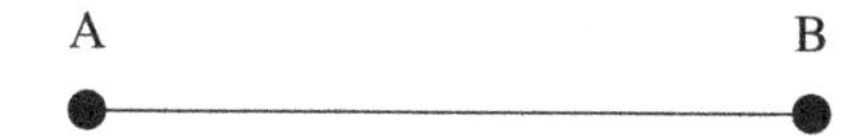

Fig. 1.4 Domain Ω defined by line AB.

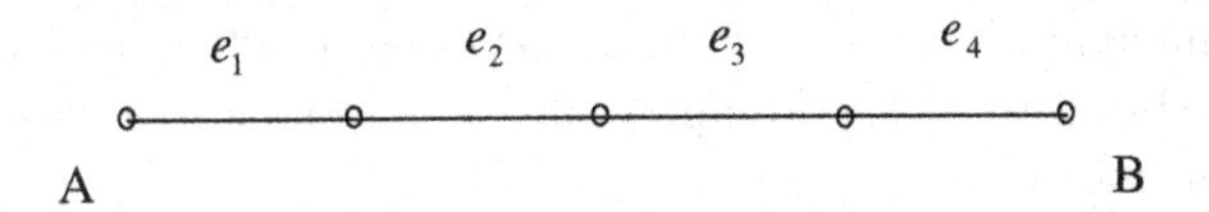

Fig. 1.5 Discretization of Ω into finite elements.

To formulate a standard Galerkin finite element solution for Eq. (1.26) we follow the steps described in the previous section.

Step 1. Discretization of the problem domain

The domain Ω is discretized into a mesh of four equal size finite elements, as shown in Fig. 1.5.

Step 2. Approximation using shape functions

Within the space of each finite element, the unknown function is approximated using interpolation functions corresponding to the two-node (linear) Lagrange elements as

$$\tilde{T} = \sum_{i=1}^{2} N_i(x) T_i \tag{1.27}$$

Where $N_i(x)$, $i = 1, 2$ are the shape functions and T_i , $i = 1, 2$ are the nodal degrees of freedom (i.e. nodal unknowns).

Step 3. Formulation of elemental weighted residual statement

The residual obtained via the insertion of $\tilde{T}$ into the differential equation is weighted and integrated over each element as

$$\int_{\Omega_e} w \left(\frac{d^2\tilde{T}}{dx^2} + \tilde{T} - 10 \right) dx = 0 \tag{1.28}$$

where w is a weighting function. In the standard Galerkin method the selected weight functions are identical to the shape functions and hence Eq. (1.28) is written as

$$\int_{\Omega_e} w_j \left[\frac{d^2 \sum_{i=1}^{2} N_i(x) T_i}{dx^2} + \sum_{i=1}^{2} N_i(x) T_i - 10 \right] dx = 0 \tag{1.29}$$

Step 4. Integration by parts (Green's theorem)

At this stage the formulated Galerkin-weighted residual equation (1.29) contains second-order derivatives. Therefore, using elements whose associated interpolation functions are only first order cannot generate an acceptable solution for this equation. To solve this difficulty the second-order derivative term in Eq. (1.29) is integrated by parts to obtain the "weak" form of the weighted residual statement as

$$\left.\frac{w_j d\sum_{i=1}^{2} N_i(x)T_i}{dx}\right|_{\Gamma_e} - \int_{\Omega_e}\left(\frac{d\sum_{i=1}^{2} N_{i(x)}T_i}{dx}\cdot\frac{dw_j}{dx}\right)dx$$

$$+\int_{\Omega_e} w_j\sum_{i=1}^{2} N_{i(x)}T_i dx - 10\int_{\Omega_e} w_j dx = 0 \tag{1.30}$$

where Γ_e represents an element boundary.

Step 5. Formulation of the elemental stiffness equations

The weight function used in the Galerkin formulation can be identical to either of the shape functions of a two-node linear element, therefore, for each weight function an equation corresponding to the weak statement (1.30) is derived.

$$\begin{cases} w_1\left.\dfrac{d(N_1T_1+N_2T_2)}{dx}\right|\Gamma_e - \displaystyle\int_{\Omega_e}\frac{dw_1}{dx}\frac{d(N_1T_1+N_2T_2)}{dx}dx \\ \quad +\displaystyle\int_{\Omega_e} w_1(N_1T_1+N_2T_2)dx - \int_{\Omega_e} 10w_1 dx = 0 \\ w_2\left.\dfrac{d(N_1T_1+N_2T_2)}{dx}\right|\Gamma_e - \displaystyle\int_{\Omega_e}\frac{dw_2}{dx}\frac{d(N_1T_1+N_2T_2)}{dx}dx \\ \quad +\displaystyle\int_{\Omega_e} w_2(N_1T_1+N_2T_2)dx - \int_{\Omega_e} 10w_2 dx = 0 \end{cases}. \tag{1.31}$$

Using matrix notation, Eq. (1.31) is written as

$$\begin{bmatrix} -\int_{\Omega_e}\left(\frac{dw_1}{dx}\frac{dN_1}{dx} - w_1N_1\right)dx & -\int_{\Omega_e}\left(\frac{dw_1}{dx}\frac{dN_2}{dx} - w_1N_2\right)dx \\ -\int_{\Omega_e}\left(\frac{dw_2}{dx}\frac{dN_1}{dx} - w_2N_1\right)dx & -\int_{\Omega_e}\left(\frac{dw_2}{dx}\frac{dN_2}{dx} - w_2N_2\right)dx \end{bmatrix}\begin{Bmatrix} T_1 \\ T_2 \end{Bmatrix}$$

$$= \begin{Bmatrix} -w_1\Phi|_{\Gamma_e} + \int_{\Omega_e} 10w_1 dx \\ -w_2\Phi|_{\Gamma_e} + \int_{\Omega_e} 10w_2 dx \end{Bmatrix} \tag{1.32}$$

In which Φ represents the boundary line term. Although the elemental stiffness equation (1.32) has a common form for all of the elements in the mesh, its utilization based on the shape functions defined in the global coordinate system is not convenient. The problem can be clearly demonstrated considering that the use of a global coordinate system results in having different interpolation functions for different elements. For example, approximate value of the field unknown anywhere inside element 1 is determined in terms of its nodal values by the following Lagrangian interpolation

$$\tilde{T}(x) = \frac{0.25 - x}{0.25} T_1 + \frac{x}{0.25} T_2$$

Therefore interpolation functions associated with element e_1 are

$$N_1 = \frac{0.25 - x}{0.25} \quad \text{and} \quad N_2 = \frac{x}{0.25}$$

However, following a similar procedure the field unknown within element 2 is interpolated as

$$\tilde{T}(x) = \frac{0.5 - x}{0.25} T_2 + \frac{x - 0.25}{0.25} T_3$$

Consequently interpolation functions corresponding to element e_2 in this example are given as

$$N_1 = \frac{0.5 - x}{0.25}, \quad N_2 = \frac{x - 0.25}{0.25}$$

and so on. Furthermore, in a global system, limits of definite integrals in the coefficient matrix will be different for each element. The described difficulties are readily resolved using a local coordinate system to define the elemental interpolation functions for a two node linear element as

$$\begin{cases} N_1 = \dfrac{\ell - x}{\ell} \\ N_2 = \dfrac{x}{\ell} \end{cases} \quad \text{which give} \quad \begin{cases} \dfrac{dN_1}{dx} = \dfrac{-1}{\ell} \\ \dfrac{dN_2}{dx} = \dfrac{1}{\ell} \end{cases}$$

where ℓ is the element length. Therefore, Eq. (1.32) is written as

$$\begin{bmatrix} -\int_0^\ell \left(\frac{dw_1}{dx}\frac{dN_1}{dx} - w_1 N_1\right) dx - \int_0^\ell \left(\frac{dw_1}{dx}\frac{dN_2}{dx} - w_1 N_2\right) dx \\ -\int_0^\ell \left(\frac{dw_2}{dx}\frac{dN_1}{dx} - w_2 N_1\right) dx - \int_0^\ell \left(\frac{dw_2}{dx}\frac{dN_2}{dx} - w_2 N_2\right) dx \end{bmatrix} \begin{Bmatrix} T_1 \\ T_2 \end{Bmatrix}$$

$$= \begin{Bmatrix} -w_1 \Phi|_{\Gamma_e} + \int_0^\ell 10 w_1 dx \\ -w_2 \Phi|_{\Gamma_e} + \int_0^\ell 10 w_2 dx \end{Bmatrix} \tag{1.33}$$

Substitution of locally defined interpolation functions into Eq. (1.33) gives

$$\frac{1}{(\ell)^2}\begin{bmatrix} \int_0^\ell(-1+\ell^2-2\ell x+x^2)dx & \int_0^\ell(1+\ell x-x^2)dx \\ \int_0^\ell(1+\ell x-x^2)dx & \int_0^\ell(-1+x^2)dx \end{bmatrix}\begin{Bmatrix} T_1 \\ T_2 \end{Bmatrix}$$

$$= \begin{Bmatrix} -\frac{\ell-x}{\ell}\Phi|_0^\ell + \int_0^\ell \frac{10}{\ell}(\ell-x)dx \\ -\frac{x}{\ell}\Phi|_0^\ell + \int_0^\ell \frac{10}{\ell}x dx \end{Bmatrix} \tag{1.34}$$

After the evaluation of the definite integrals in the coefficient matrix and the boundary line terms in the right-hand side, Eq. (1.34) gives

$$\begin{bmatrix} \frac{\ell}{3}-\frac{1}{\ell} & \frac{1}{\ell}+\frac{\ell}{6} \\ \frac{1}{\ell}+\frac{\ell}{6} & \frac{\ell}{3}-\frac{1}{\ell} \end{bmatrix}\begin{Bmatrix} T_1 \\ T_2 \end{Bmatrix} = \begin{Bmatrix} q_1+5\ell \\ -q_2+5\ell \end{Bmatrix} \tag{1.35}$$

Therefore using the discretization shown in Fig. 1.5 where $\ell = 0.25$

$$\begin{bmatrix} -\frac{11.75}{3} & \frac{24.25}{6} \\ \frac{24.25}{6} & -\frac{11.75}{3} \end{bmatrix}\begin{Bmatrix} T_1 \\ T_2 \end{Bmatrix} = \begin{Bmatrix} q_1+1.25 \\ -q_2+1.25 \end{Bmatrix} \tag{1.36}$$

Equation (1.36) is the common form of the elemental stiffness equation in this example.

Step 6. Assembly of the elemental stiffness equations into a global system of algebraic equations

Elemental stiffness equations are assembled over their common nodes to yield

$$\begin{bmatrix} -\frac{11.75}{3} & \frac{24.25}{6} & 0 & 0 & 0 \\ \frac{24.25}{6} & -\frac{23.5}{3} & \frac{24.25}{6} & 0 & 0 \\ 0 & \frac{24.25}{6} & -\frac{23.5}{3} & \frac{24.25}{6} & 0 \\ 0 & 0 & \frac{24.25}{6} & -\frac{23.5}{3} & \frac{24.25}{6} \\ 0 & 0 & 0 & \frac{24.25}{6} & -\frac{11.75}{3} \end{bmatrix}\begin{Bmatrix} T_1 \\ T_2 \\ T_3 \\ T_4 \\ T_5 \end{Bmatrix} = \begin{Bmatrix} q_1+1.25 \\ 2.5 \\ 2.5 \\ 2.5 \\ -q_5+1.25 \end{Bmatrix} \tag{1.37}$$

Note that in equation system (1.37) the coefficients matrix is symmetric, sparse (i.e. a significant number of its members are zero) and banded. The symmetry of coefficients matrix in the global finite element equations depends on the nature of the original differential equation being solved and is not guaranteed for all cases (in particular, in most fluid flow problems this matrix will not be symmetric a). However, finite element formulations always yield sparse and banded sets of equations. This property is usually utilized to minimize computing costs in complex problems.

Step 7. Imposition of the boundary conditions

Prescribed values of the unknown function at the boundaries of Ω (i.e. $T_1 = 0$, $T_5 = 1$) are inserted into the system of algebraic equations (1.37) and redundant equations corresponding to the boundary nodes eliminated from the set. After algebraic manipulations the following set of equations is obtained

$$\begin{bmatrix} -\frac{23.5}{3} & \frac{24.25}{6} & 0 \\ \frac{24.25}{6} & -\frac{23.5}{3} & \frac{24.25}{6} \\ 0 & \frac{24.25}{6} & -\frac{23.5}{3} \end{bmatrix} \begin{Bmatrix} T_2 \\ T_3 \\ T_4 \end{Bmatrix} = \begin{Bmatrix} 2.5 - \frac{24.25}{6} \times 1 \\ 2.5 \\ 2.5 - \frac{24.25}{6} \times 4 \end{Bmatrix} \tag{1.38}$$

Step 8. Solution of the algebraic equations

Equation set (1.38) is a determinate system and its solution gives

$$\begin{cases} T_2 = 0.95 \\ T_3 = 1.4598 \\ T_4 = 2.4979 \end{cases} \tag{1.39}$$

The analytical solution for this problem is

$$T(x) = -9\cos(x) - 1.3515\sin(x) + 10 \tag{1.40}$$

The comparison of the discrete solutions given in set (1.39) with their corresponding values from Eq. (1.40) reveals an acceptable degree of accuracy in the finite element results (Fig. 1.6), despite using a coarse mesh of only four elements. For example, the relative error corresponding to the node located at the middle of the domain is less than 0.1%. A further point to consider is that in this simple one-dimensional example coordinate mapping to a master element is not used.

However, as mentioned earlier, in general, a mapping from the global domain to a master element should be carried out to have a uniform limit for the integrals in the elemental matrices.

In the finite element solution of a realistic engineering problem, the global set of algebraic equations obtained after the assembly of elemental contributions will be very large (usually consisting of several thousand algebraic equations). In many cases the global equations needed to be solved are also ill-conditioned (Gerald and Wheatley, 1984). Ill-conditioned sets usually include equations in which the coefficients of unknowns range over many orders of magnitude. If procedures which cannot preserve the precision of the calculations are used, round-off errors badly affect the small

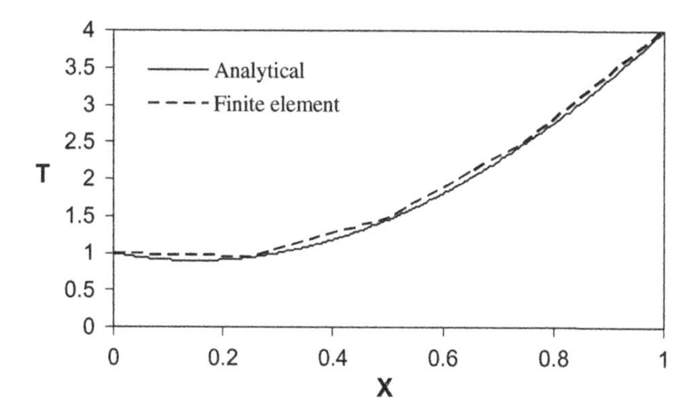

Fig. 1.6 Comparison between exact and finite element solutions.

coefficients and hence the nature of equations changes. This results in a very high degree of approximation, rendering the solutions unacceptable. Therefore, the solution of the global system of algebraic equations is regarded as one of the most important steps in the finite element modeling of realistic problems. Commonly used techniques for the solution of sets of algebraic equations in finite element schemes include "Direct" elimination methods, such as the "Frontal Solution" or "LU Decomposition," and "Iterative" procedures, such as the "Preconditioned Conjugate Gradient Method." Computing economy, speed, and the required accuracy of the solutions are the most important factors that should be taken into account in selecting solver routines for finite element programming. Description and full listings of computer codes based on these techniques can be found in the literature (e.g. see Hood, 1976; Hinton and Owen, 1977; Irons and Ahmad, 1980).

1.3.2 *Example 2*

To consolidate the description of the weighted residuals finite element method, we now consider the solution of the following two-dimensional Poisson equation over a rectangular domain subject to homogeneous boundary conditions (i.e. $T = 0$ everywhere on the domain boundary).

$$\frac{\partial^2 T}{\partial x^2} + \frac{\partial^2 T}{\partial y^2} = 2(x^2 + y^2 - x - y) \tag{1.41}$$

Starting with the usual step in the Galerkin finite element method, the problem domain is discretized into four rectangular elements as shown in Fig. 1.7.

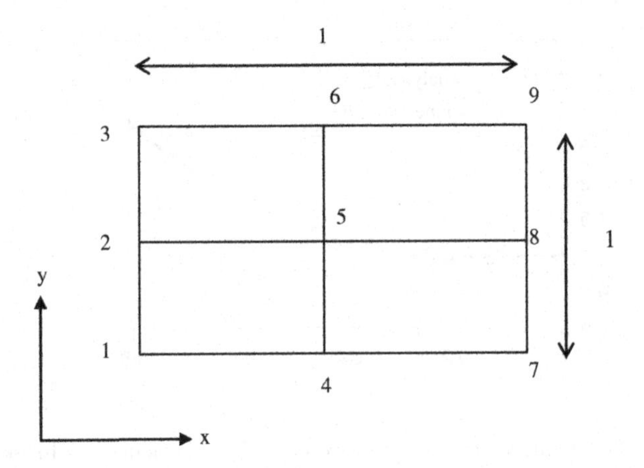

Fig. 1.7 Discretized domain showing a mesh of four bilinear elements.

Within each element the unknown function T is approximated using interpolation functions constructed as the products of linear Lagrangian interpolation functions in x and y directions. Therefore these elements are said to be four-node bilinear tensor product elements. A detailed explanation of the derivation of these functions are given in the next chapter of this book.

$$T \approx \tilde{T} = \sum_{i=1}^{4} N_i(x,y).T_i \tag{1.42}$$

where $N_i(x,y)$ are the interpolation functions associated with nodes $i = 1,\ldots,4$. The weighted residual statement of the problem is therefore written as

$$\iint_{\Omega_e} w\left(\frac{\partial^2\tilde{T}}{\partial x^2} + \frac{\partial^2\tilde{T}}{\partial y^2} - 2(x^2+y^2-x-y)\right)dxdy = 0 \tag{1.43}$$

In the standard Galerkin method, weight functions are identical to the shape functions and hence

$$\iint_{\Omega_e} [N]^T\left(\frac{\partial^2\tilde{T}}{\partial x^2} + \frac{\partial^2\tilde{T}}{\partial y^2} - 2x^2 - 2y^2 + 2x + 2y\right)dxdy = 0 \tag{1.44}$$

For simplicity of writing we write Eq. (1.44) as

$$\int_A [N]^T\left(\left(\frac{\partial^2\tilde{T}}{\partial x^2} + \frac{\partial^2\tilde{T}}{\partial y^2} - 2(x^2+y^2-x-y)\right)dA = 0 \tag{1.45}$$

After the application of Green's theorem second-order derivatives disappear and we have

$$\int_{\Gamma} [N]^T \left(\frac{\partial \tilde{T}}{\partial x} \cos\theta + \frac{\partial \tilde{T}}{\partial y} \sin\theta \right) d\Gamma - \int_A \left(\frac{\partial [N]^T}{\partial x} \frac{\partial \tilde{T}}{\partial x} + \frac{\partial [N]^T}{\partial y} \frac{\partial \tilde{T}}{\partial y} \right) dA$$

$$- \int_A [N]^T (2(x^2 + y^2 - x - y)) dA = 0 \qquad (1.46)$$

where Γ is the element boundary line and θ is the angle between outward normal vector to the boundary and Γ. The weighted residual statement (1.46) consists of the following three components

$$[k^{(e)}] = \left(\int_A \left(\frac{\partial [N]^T}{\partial x} \frac{\partial \tilde{T}}{\partial x} + \frac{\partial [N]^T}{\partial y} \frac{\partial \tilde{T}}{\partial y} \right) \right) dA \qquad (1.47)$$

$$\{f^{(e)}\} = \int_A [N]^T (2(x^2 + y^2 - x - y)) dA \qquad (1.48)$$

$$\{I^{(e)}\} = \int_{\Gamma} [N]^T \left(\frac{\partial \tilde{T}}{\partial x} \cos\theta + \frac{\partial \tilde{T}}{\partial y} \sin\theta \right) d\Gamma \qquad (1.49)$$

We now calculate each of these terms using the previously described mapping onto a master element with its associated local coordinate system shown in Fig. 1.8

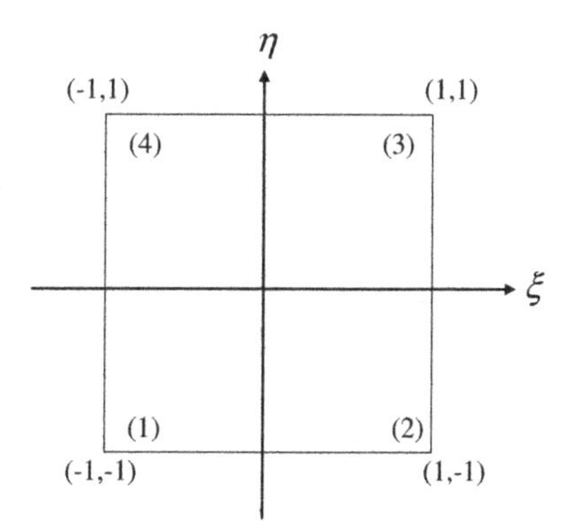

Fig. 1.8 Four-node square element and its associated local coordinate system (ξ, η).

$$[k^{(e)}] = \frac{1}{16}\int_{-1}^{+1}\int_{-1}^{+1}\begin{bmatrix} \frac{\partial N_i}{\partial \xi} & \frac{\partial N_i}{\partial \eta} \\ \frac{\partial N_j}{\partial \xi} & \frac{\partial N_j}{\partial \eta} \\ \frac{\partial N_k}{\partial \varepsilon} & \frac{\partial N_k}{\partial \eta} \\ \frac{\partial N_l}{\partial \xi} & \frac{\partial N_l}{\partial \eta} \end{bmatrix} \begin{bmatrix} \frac{\partial N_i}{\partial \xi} & \frac{\partial N_j}{\partial \xi} & \frac{\partial N_k}{\partial \xi} & \frac{\partial N_l}{\partial \xi} \\ \frac{\partial N_i}{\partial \eta} & \frac{\partial N_j}{\partial \eta} & \frac{\partial N_k}{\partial \eta} & \frac{\partial N_l}{\partial \eta} \end{bmatrix} d\xi d\eta \quad (1.50)$$

where i, j, k, and l are node indices. Substitution for the derivatives of the shape functions in Eq. (1.50) gives

$$[k^{(e)}] = \frac{1}{16}\int_{-1}^{+1}\int_{-1}^{+1}\begin{bmatrix} -(1-\eta) & -(1-\xi) \\ (1-\eta) & -(1+\xi) \\ (1+\eta) & (1+\xi) \\ -(1+\eta) & (1-\xi) \end{bmatrix}$$

$$\begin{bmatrix} (1-\eta) & (1-\eta) & (1+\eta) & -(1+\eta) \\ -(1-\xi) & -(1+\xi) & (1+\xi) & (1-\xi) \end{bmatrix} d\xi d\eta \quad (1.51)$$

and

$$[k^{(e)}] = \frac{1}{16}\int_{-1}^{+1}\int_{-1}^{+1}\left[\begin{matrix} (1-\eta)^2 + (1-\xi)^2 & (1-\xi^2) - (1-\eta)^2 \\ (1-\xi^2) - (1-\eta)^2 & (1-\eta)^2 + (1+\xi)^2 \\ -(1-\eta^2) - (1-\xi^2) & (1-\eta^2) - (1+\xi)^2 \\ (1-\eta^2) - (1-\xi)^2 & -(1-\eta)^2 - (1-\xi^2) \end{matrix}\right.$$

$$\left.\begin{matrix} -(1-\eta^2) - (1-\xi^2) & (1-\eta^2) - (1-\xi)^2 \\ (1-\eta^2) - (1+\xi)^2 & -(1-\eta^2) - (1-\xi^2) \\ (1+\eta)^2 + (1+\xi)^2 & (1-\xi^2) - (1+\eta)^2 \\ (1-\xi^2) - (1+\eta)^2 & (1+\eta)^2 + (1-\xi)^2 \end{matrix}\right] d\xi \quad (1.52)$$

After the evaluation of the integrals, Eq. (1.52) gives

$$[k^{(e)}] = \begin{bmatrix} \overset{i}{\frac{2}{3}} & \overset{j}{-\frac{1}{6}} & \overset{k}{-\frac{1}{3}} & \overset{l}{-\frac{1}{6}} \\ -\frac{1}{6} & \frac{2}{3} & -\frac{1}{6} & -\frac{1}{3} \\ -\frac{1}{3} & -\frac{1}{6} & \frac{2}{3} & -\frac{1}{6} \\ -\frac{1}{6} & -\frac{1}{3} & -\frac{1}{6} & \frac{2}{3} \end{bmatrix} \begin{matrix} i \\ j \\ k \\ l \end{matrix} \quad (1.53)$$

Similarly

$$\{f^{(e)}\} = \int_A [N]^T \left(2\left(\frac{(1+\xi)^2}{16} + \frac{(1+\eta)^2}{16} - \frac{(1+\xi)}{4} - \frac{(1+\eta)}{4}\right)\right) dA \tag{1.54}$$

Hence we have

$$\{f^{(e)}\} = \frac{1}{32}\int_{-1}^{+1}\int_{-1}^{+1}\left(\frac{(1+\xi)^2}{16} + \frac{(1+\eta)^2}{16} - \frac{(1+\xi)}{4} - \frac{(1+\eta)}{4}\right)$$
$$\begin{bmatrix}(1-\xi)(1-\eta)\\(1+\xi)(1-\eta)\\(1+\xi)(1+\eta)\\(1-\xi)(1+\eta)\end{bmatrix} d\xi d\eta \tag{1.55}$$

or

$$\{f^{(e)}\} = \frac{1}{32}\int_{-1}^{+1}\int_{-1}^{+1}$$
$$\begin{bmatrix}\frac{(1+\xi)^2(1-\xi)(1-\eta)}{16} + \frac{(1+\eta)^2(1-\eta)(1-\xi)}{16} - \frac{(1-\xi)^2(1-\eta)}{4} - \frac{(1-\eta)^2(1-\xi)}{4}\\ \frac{(1+\xi)^3(1-\eta)}{16} + \frac{(1+\eta)^2(1-\eta)(1+\xi)}{16} - \frac{(1+\xi)^2(1-\eta)}{4} - \frac{(1-\eta)^2(1+\xi)}{4}\\ \frac{(1+\xi)^3(1+\eta)}{16} + \frac{(1+\eta)^2(1+\xi)}{16} - \frac{(1+\xi)^2(1+\eta)}{4} - \frac{(1+\eta)^2(1+\xi)}{4}\\ \frac{(1+\xi)^2(1-\xi)(1-\eta)}{16} + \frac{(1+\eta)^2(1-\eta)(1-\xi)}{16} - \frac{(1-\xi)^2(1-\eta)}{4} - \frac{(1-\eta)^2(1-\xi)}{4}\end{bmatrix} d\xi d\eta \tag{1.56}$$

and finally

$$\{f^{(e)}\} = \begin{bmatrix}-\frac{1}{32}\\-\frac{1}{24}\\-\frac{5}{96}\\-\frac{1}{24}\end{bmatrix} \begin{matrix}i\\j\\k\\l\end{matrix} \tag{1.57}$$

The flux term along the inner boundaries between the elements of the domain cancel each other out during the assembly of the elemental stiffness equations. As in the exterior boundaries, essential boundary conditions (i.e. function values) are given there is no need to solve any equations. Therefore there is no need to calculate the third component of Eq. (1.46). Assembling the elemental equations we obtain the following global stiffness equation and load vectors (i.e. R.H.S.) for the present example.

Global stiffness equation

$$\begin{bmatrix} \overset{1}{0.66670} & \overset{2}{-0.1667} & \overset{3}{0} & \overset{4}{-0.1667} & \overset{5}{-0.3333} & \overset{6}{0} & \overset{7}{0} & \overset{8}{0} & \overset{9}{0} \\ -0.1667 & 1.33330 & -0.1667 & -0.3333 & -0.3333 & -0.3333 & 0 & 0 & 0 \\ 0 & -0.1667 & 0.66670 & 0 & -0.3333 & -0.1667 & 0 & 0 & 0 \\ -0.1667 & -0.3333 & 0 & 1.33330 & -0.3333 & 0 & -0.1667 & -0.3333 & 0 \\ -0.3333 & -0.3333 & -0.3333 & -0.3333 & 2.66670 & -0.3333 & -0.3333 & -0.3333 & -0.3333 \\ 0 & -0.3333 & -0.1667 & 0 & -0.3333 & 1.33330 & 0 & -0.3333 & -0.1667 \\ 0 & 0 & 0 & -0.1667 & -0.3333 & 0 & 0.66670 & -0.1667 & 0 \\ 0 & 0 & 0 & -0.3333 & -0.3333 & -0.3333 & -0.1667 & 1.33330 & -0.1667 \\ 0 & 0 & 0 & 0 & -0.3333 & -0.1667 & 0 & -0.1667 & 0.66670 \end{bmatrix} \tag{1.58}$$

Load vector

$$\begin{bmatrix} -0.0313 \\ -0.0729 \\ -0.0417 \\ -0.0729 \\ -0.1667 \\ -0.0938 \\ -0.0417 \\ -0.0938 \\ -0.0521 \end{bmatrix} . \tag{1.59}$$

After the imposition of the boundary conditions and simplifying the global system of equations we obtain

$$[2.66670][T_5] = 0.1667 \tag{1.60}$$

and

$$T_5 = 6.25e^{-02} \tag{1.61}$$

The analytical solution of the original equation is

$$T = x(1-x) \quad y(1-y) \tag{1.62}$$

Insertion of the coordinates at the centre of the domain gives the exact result for this point again as $T = 6.25e^{-02}$. Therefore despite using a very coarse mesh the application of standard Galerkin method in this example yields a super-convergent result.

References

[1] Dhatt, G. and Touzot, G., *The Finite Element Displayed*, John Wiley & Sons Inc., 1985.
[2] Gerald, C.F. and Wheatley, P.O., *Applied Numerical Analysis*, 3rd ed., Addison-Wesley, Reading, Massachusetts, 1984.
[3] Hinton, E. and Owen, D.J.R., *Finite Element Programming*, Academic Press, 1977.
[4] Hood, P., Frontal solution program for unsymmetric matrices, *Int. J. Numer. Meth. Eng.*, 1976; 10; 379–399.
[5] Hughes, T.J.R. and Brooks, A.N., A multidimensional upwind scheme with no cross-wind diffusion, in *Finite Element Methods for Convection Dominated Flows*, AMD Vol. 34, TJR Hughes (Ed.), ASME, New York, 1979.
[6] Irons, B. and Ahmad, S., *Techniques of Finite Elements*, Ellis Horwood/John Wiley & Sons, 1980.
[7] Nassehi, V., *Practical Aspects of Finite Element Modelling of Polymer Processing*, John Wiley and Sons Ltd, Chichester, 2002.
[8] Petera, J., Nassehi, V. and Pittman, J.F.T, Petrov–Galerkin methods on isoparametric bilinear and biquadratic elements tested for a scalar convection-diffusion problem, *Int. J. Numer. Meth. Heat Fluid Flow*, 1993; 3; 205–222.
[9] Spiegel, M.R., *Vector Analysis, Schaum's Outline Series*, McGraw-Hill Book Company, New York, 1974.
[10] Zienkiewicz, O.C. and Morgan, K., *Finite Elements and Approximation*, John Wiley & Sons, New York, 1983.
[11] Zienkiewicz, O.C. and Taylor, R.L., *The Finite Element Method*, 4th ed., Vol. 1 and 2, McGraw-Hill, London, 1994.

CHAPTER 2

Shape Functions and Fundamental Properties of Finite Elements

In this chapter the most important methods used to derive shape functions of common types of elements are explained. It is shown that basic properties of finite elements and the approximations generated by them depend on the nature of their associated shape functions. Traditionally, shape functions associated with a finite element are based on the interpolation model which can be used to formulate an approximation for an unknown for a field problem within the space of a finite element. After the explanation of the basic procedures used to derive shape functions of common types of elements, a number of benchmark problems are solved. These examples are chosen to illustrate the shortcomings of common types of finite elements and their influence on the accuracy and reliability of various types of weighted residuals finite element schemes. Such considerations lead to the explanation of techniques used to enhance the accuracy and stability of the finite element schemes and the introduction of the basic concept of enriched elements. As described in the following chapters of this book, enriched elements provide very reliable techniques for the accurate finite element solution of multiscale problems whilst maintaining computing economy.

In Chap. 1 construction of elemental approximations for field unknowns using low-order interpolation models is explained. Accuracy of such approximations can logically be expected to improve by increasing the order of the interpolation polynomials, that is, using higher-order elements. We now consider this concept in more detail.

Finite elements can be classified into different types such as simplex, complex, and multiplex families. Simplex elements are those in which the interpolation is based on complete linear polynomials. The best example of a simplex element is a triangle with three nodes located on its vertices. Elements associated with complete order quadratic, cubic or higher-order polynomials are called complex elements. Elements which are associated with incomplete polynomials are called multiplex elements. Geometrically, simplex and complex elements have triangular or tetrahedral shapes but the multiplex elements are quadrilaterals or their three-dimensional equivalents (Rao, 1989).

2.1 Interpolation Polynomials

In one dimension the complete nth order polynomial can be written as (Huebner *et al.*, 2001)

$$T(x) = \sum_{k=1}^{n+1} a_k x^i, \quad i \leq n. \tag{2.1}$$

For example, in one-dimensional cases we have

$$\begin{aligned} T(x) &= a_1 + a_2 x && \text{for } n = 1 \text{ (linear element)}, \\ T(x) &= a_1 + a_2 x + a_3 x^2 && \text{for } n = 2 \text{ (quadratic element)}, \\ T(x) &= a_1 + a_2 x + a_3 x^2 + a_4 x^3 && \text{for } n = 3 \text{ (cubic element)}. \end{aligned} \tag{2.2}$$

In two dimensions the complete nth order polynomial can be written as

$$T(x) = \sum_{k=1}^{\{0.5(n+1)(n+2)\}} a_k x^i y^j, \quad i + j \leq n. \tag{2.3}$$

Therefore

$$\begin{cases} T(x) = a_1 + a_2 x + a_3 y & \text{for } n = 1 \text{ (three-noded triangle)} \\ T(x) = a_1 + a_2 x + a_3 y + a_4 xy & \\ \qquad + a_5 x^2 + a_6 y^2 & \text{for } n = 2 \text{ (six-noded triangle)} \end{cases} \tag{2.4}$$

and so on.

In three dimensions the complete nth order polynomial can be written as

$$T(x) = \sum_{k=1}^{\{(n+1)(n+2)(n+3)/6\}} a_k x^i y^j z^l, \quad i + j + l \leq n. \tag{2.5}$$

Therefore

$$\begin{aligned} T(x) &= a_1 + a_2 x + a_3 y + a_4 z && \text{for } n = 1, \\ T(x) &= a_1 + a_2 x + a_3 y + a_4 z + a_5 xy + a_6 xz && \\ &\quad + a_7 yz + a_8 x^2 + a_9 y^2 + a_{10} z^2 && \text{for } n = 2, \end{aligned} \tag{2.6}$$

Here $n = 1$ gives a four-noded tetrahedral element and $n = 2$ gives a ten-noded tetrahedral element.

2.2 Shape Functions of Simplex Elements

We now consider the derivation of interpolation (shape) functions for one-, two- and three-dimensional simplex elements. These elements are probably the most widely used finite elements in the solution of engineering problems.

2.2.1 *One-dimensional linear element*

Consider a one-dimensional element (line segment) of length l as shown in Fig. 2.1.

The field variable T inside the element is approximated using linear interpolation as

$$\tilde{T} = ax + b. \tag{2.7}$$

The nodal values of the field variable are

$$\begin{cases} T_1 = ax_1 + b \\ T_2 = ax_2 + b \end{cases}. \tag{2.8}$$

In the matrix form we have (Nassehi, 2002)

$$\begin{bmatrix} T_1 \\ T_2 \end{bmatrix} = \begin{bmatrix} 1 & x_1 \\ 1 & x_2 \end{bmatrix} \begin{bmatrix} a \\ b \end{bmatrix} \tag{2.9}$$

or

$$T^e = CA, \tag{2.10}$$

$$\tilde{T} = PA = PC^{-1}T^e, \tag{2.11}$$

where $P = [1 \quad x]$,

$$\tilde{T} = NT^e. \tag{2.12}$$

Where N is the set of shape functions derived using the outlined interpolation procedure

$$N = PC^{-1} = [1 \quad x] \begin{bmatrix} 1 & x_1 \\ 1 & x_2 \end{bmatrix}^{-1} = \frac{1}{x_2 - x_1} [1 \quad x] \begin{bmatrix} x_2 & -x_1 \\ -1 & 1 \end{bmatrix}, \tag{2.13}$$

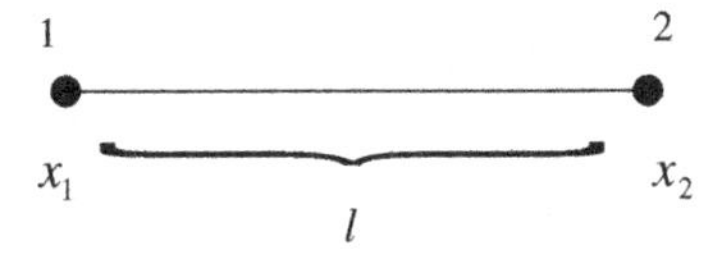

Fig. 2.1 One-dimensional linear element.

$$
\begin{aligned}
N &= \frac{1}{x_2 - x_1}\left[x_2 - x,\ -x_1 + x\right], \\
x_2 &= l, \quad x_1 = 0, \\
N &= \frac{1}{l}\left[l - x,\ x\right],
\end{aligned} \tag{2.14}
$$

$$
N_1 = \frac{l - x}{l}, \quad N_2 = \frac{x}{l}. \tag{2.15}
$$

The relation between local coordinate systems $x(0, l)$ and $\xi(-1, +1)$ is

$$
x = \frac{l}{2}(1 + \xi). \tag{2.16}
$$

In the normalized local coordinate system of $\xi(-1, +1)$ we have

$$
N_1 = \frac{1}{2}(1 - \xi), \quad N_2 = \frac{1}{2}(1 + \xi). \tag{2.17}
$$

2.2.2 *Two-dimensional simplex element*

The two-dimensional simplex element is a triangle with three nodes located at its vertices (Fig. 2.2). The field variable inside the element is approximated as

$$
\tilde{T} = a + bx + cy \tag{2.18}
$$

At nodes 1, 2, and 3 we have

$$
\begin{cases}
T_1 = a + bx_1 + cy_1 \\
T_2 = a + bx_2 + cy_2 \\
T_3 = a + bx_3 + cy_3
\end{cases}. \tag{2.19}
$$

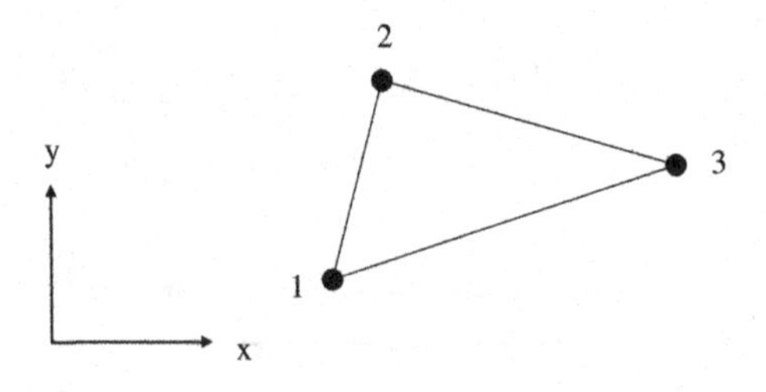

Fig. 2.2 A three-node triangular element.

Using matrix form

$$\begin{bmatrix} 1 & x_1 & y_1 \\ 1 & x_2 & y_2 \\ 1 & x_3 & y_3 \end{bmatrix} \begin{bmatrix} a \\ b \\ c \end{bmatrix} = \begin{bmatrix} T_1 \\ T_2 \\ T_3 \end{bmatrix}. \tag{2.20}$$

The inverse of the coefficient matrix C is found as

$$C^{-1} = \frac{1}{2|C|} \begin{bmatrix} \begin{vmatrix} x_2 & y_2 \\ x_3 & y_3 \end{vmatrix} & -\begin{vmatrix} 1 & y_2 \\ 1 & y_3 \end{vmatrix} & \begin{vmatrix} 1 & x_2 \\ 1 & x_3 \end{vmatrix} \\ -\begin{vmatrix} x_1 & y_1 \\ x_3 & y_3 \end{vmatrix} & \begin{vmatrix} 1 & y_1 \\ 1 & y_3 \end{vmatrix} & -\begin{vmatrix} 1 & x_1 \\ 1 & x_3 \end{vmatrix} \\ \begin{vmatrix} x_1 & y_1 \\ x_2 & y_2 \end{vmatrix} & -\begin{vmatrix} 1 & y_1 \\ 1 & y_2 \end{vmatrix} & \begin{vmatrix} 1 & x_1 \\ 1 & x_2 \end{vmatrix} \end{bmatrix}^T \tag{2.21}$$

$$= \frac{1}{2|C|} \begin{bmatrix} x_2y_3 - x_3y_2 & y_2 - y_3 & x_3 - x_2 \\ x_3y_1 - x_1y_3 & y_3 - y_1 & x_1 - x_3 \\ x_1y_2 - x_2y_1 & y_1 - y_2 & x_2 - x_1 \end{bmatrix}^T .$$

This operation can be simplified if instead of a global system we consider the following "master" element with its local coordinate system as shown in Fig. 2.3.

In this element node 1 is the origin of the coordinate system and nodes 2 and 3 lay on the x and y axes, respectively. The shape functions in this case are found as

$$N = PC^{-1} = \frac{1}{|C|} [1 \quad x \quad y] \begin{bmatrix} x_2y_3 & 0 & 0 \\ -y_3 & 0 & 0 \\ -x_2 & 0 & x_2 \end{bmatrix}, \tag{2.22}$$

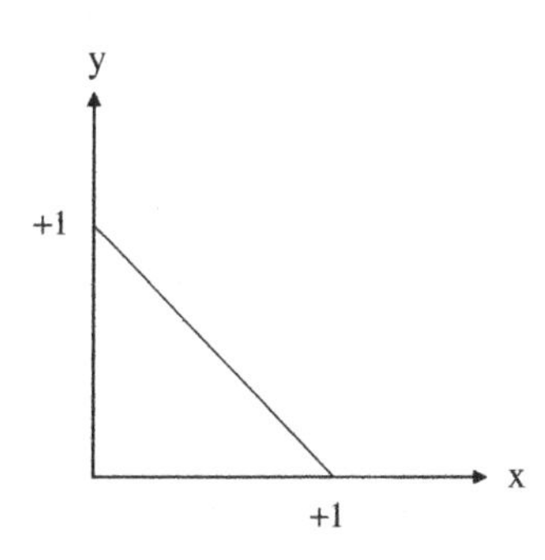

Fig. 2.3 A triangular element with its associated local coordinate system $x(0, +1)$ and $y(0, +1)$.

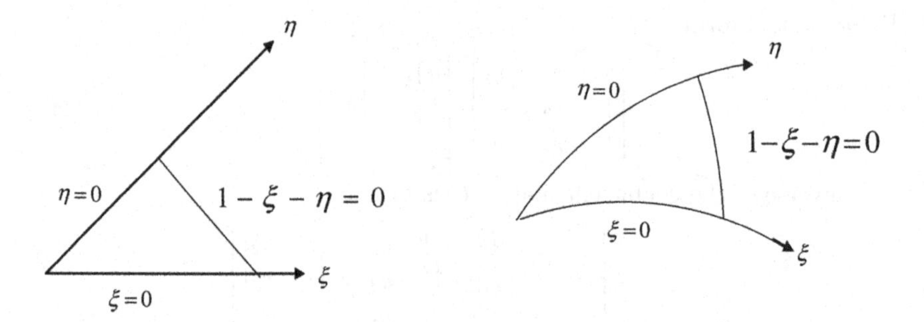

Fig. 2.4 Non-orthogonal and curvilinear local coordinates.

and

$$|C| = x_2y_3 - x_3y_2 = x_2y_3 = 2s, \tag{2.23}$$

where s is the surface area of the triangular element. Local coordinate system does not need to be orthogonal and can also be chosen to be curvilinear with axis defined as (ξ, η) (Fig. 2.4).

As mentioned in Chap. 1, ordinary elements in a global mesh can be mapped into a master element defined in terms of its local coordinates. A system of local natural coordinates based on previously mentioned area coordinates can also be used. However, in this book we have used the triangular simplex element shown in Fig. 2.3 with coordinates defined as

$$x_1 = 0, \quad y_1 = 0, \quad x_3 = 0, \quad \text{and} \quad y_2 = 0. \tag{2.24}$$

Therefore the shape functions associated with this element are derived as

$$\begin{cases} N_1 = 1 - \dfrac{x}{x_2} - \dfrac{y}{y_3} \\ N_2 = \dfrac{x}{x_2} \\ N_3 = \dfrac{y}{y_3} \end{cases}. \tag{2.25}$$

The coordinate system of (ξ, η) can be related to the (x, y) system as

$$\xi = \frac{x}{x_2}, \quad \eta = \frac{y}{y_3}, \tag{2.26}$$

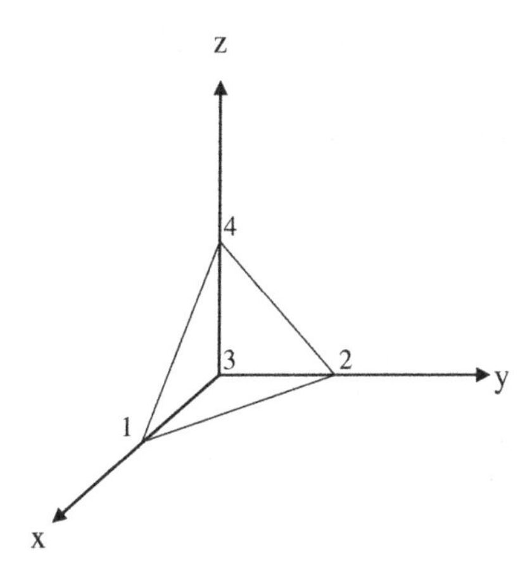

Fig. 2.5 A master tetrahedral element with its local coordinate system (x, y, z).

and replacing from relations (2.26) into Eq. (2.25)

$$\begin{cases} N_1 = 1 - \xi - \eta \\ N_2 = \xi \\ N_3 = \eta \end{cases} . \tag{2.27}$$

2.2.3 *Three-dimensional simplex element*

To derive the shape functions of three-dimensional simplex elements, i.e. the four-noded tetrahedron shown in Fig. 2.5 we follow a similar procedure. The nodes of this element are located at its vertices and with respect to the adopted coordinate system we have

$$x_3 = y_3 = z_3 = 0, \quad x_2 = z_2 = 0, \quad y_1 = z_1 = 0, \quad x_4 = y_4 = 0. \tag{2.28}$$

The field variable inside the element is approximated as

$$\tilde{T} = a + bx + cy + dz. \tag{2.29}$$

Using matrix notation we have

$$\begin{bmatrix} T_1 \\ T_2 \\ T_3 \\ T_4 \end{bmatrix} = \begin{bmatrix} 1 & x_1 & y_1 & z_1 \\ 1 & x_2 & y_2 & z_2 \\ 1 & x_3 & y_3 & z_3 \\ 1 & x_4 & y_4 & z_4 \end{bmatrix} \begin{bmatrix} a \\ b \\ c \\ d \end{bmatrix} . \tag{2.30}$$

The inverse of the coefficient matrix C is found as

$$C^{-1} = \frac{1}{|C|} \begin{bmatrix} \begin{vmatrix} x_2 & y_2 & z_2 \\ x_3 & y_3 & z_3 \\ x_4 & y_4 & z_4 \end{vmatrix} & -\begin{vmatrix} 1 & y_2 & z_2 \\ 1 & y_3 & z_3 \\ 1 & y_4 & z_4 \end{vmatrix} & \begin{vmatrix} 1 & x_2 & z_2 \\ 1 & x_3 & z_3 \\ 1 & x_4 & z_4 \end{vmatrix} & -\begin{vmatrix} 1 & x_2 & y_2 \\ 1 & x_3 & y_3 \\ 1 & x_4 & y_4 \end{vmatrix} \\ -\begin{vmatrix} x_1 & y_1 & z_1 \\ x_3 & y_3 & z_3 \\ x_4 & y_4 & z_4 \end{vmatrix} & \begin{vmatrix} 1 & y_1 & z_1 \\ 1 & y_3 & z_3 \\ 1 & y_4 & z_4 \end{vmatrix} & -\begin{vmatrix} 1 & x_1 & z_1 \\ 1 & x_3 & z_3 \\ 1 & x_4 & z_4 \end{vmatrix} & \begin{vmatrix} 1 & x_1 & y_1 \\ 1 & x_3 & y_3 \\ 1 & x_4 & y_4 \end{vmatrix} \\ \begin{vmatrix} x_1 & y_1 & z_1 \\ x_2 & y_2 & z_2 \\ x_4 & y_4 & z_4 \end{vmatrix} & -\begin{vmatrix} 1 & y_1 & z_1 \\ 1 & y_2 & z_2 \\ 1 & y_4 & z_4 \end{vmatrix} & \begin{vmatrix} 1 & x_1 & z_1 \\ 1 & x_2 & z_2 \\ 1 & x_4 & z_4 \end{vmatrix} & -\begin{vmatrix} 1 & x_1 & y_1 \\ 1 & x_2 & y_2 \\ 1 & x_4 & y_4 \end{vmatrix} \\ -\begin{vmatrix} x_1 & y_1 & z_1 \\ x_2 & y_2 & z_2 \\ x_3 & y_3 & z_3 \end{vmatrix} & \begin{vmatrix} 1 & y_1 & z_1 \\ 1 & y_2 & z_2 \\ 1 & y_3 & z_3 \end{vmatrix} & -\begin{vmatrix} 1 & x_1 & z_1 \\ 1 & x_2 & z_2 \\ 1 & x_3 & z_3 \end{vmatrix} & \begin{vmatrix} 1 & x_1 & y_1 \\ 1 & x_2 & y_2 \\ 1 & x_3 & y_3 \end{vmatrix} \end{bmatrix}^T . \tag{2.31}$$

Replacing for the known coordinate values from Eq. (2.28)

$$C^{-1} = \frac{1}{|C|} \begin{bmatrix} 0 & y_2 z_4 & 0 & 0 \\ 0 & 0 & x_1 z_4 & 0 \\ x_1 y_2 z_4 & -y_2 z_4 & -x_1 z_4 & -x_1 y_2 \\ 0 & 0 & 0 & x_1 y_2 \end{bmatrix}^T \tag{2.32}$$

and

$$|C| = x_1 y_2 z_4. \tag{2.33}$$

Therefore

$$N = PC^{-1} = \frac{1}{|C|} [1 \quad x \quad y \quad z] \begin{bmatrix} 0 & 0 & x_1 y_2 z_4 & 0 \\ y_2 z_4 & 0 & -y_2 z_4 & 0 \\ 0 & x_1 z_4 & -x_1 z_4 & 0 \\ 0 & 0 & -x_1 y_2 & x_1 y_2 \end{bmatrix}, \tag{2.34}$$

and finally

$$\begin{aligned} N_1 &= \frac{1}{|C|} y_2 z_4 x = \frac{x}{x_1}, \\ N_2 &= \frac{1}{|C|} x_1 z_4 y = \frac{y}{y_2}, \\ N_3 &= \frac{1}{|C|} (x_1 y_2 z_4 - y_2 z_4 x - x_1 z_4 y - x_1 y_2 z) \\ &= 1 - \frac{x}{x_1} - \frac{y}{y_2} - \frac{z}{z_4}, \\ N_4 &= \frac{1}{|C|} x_1 y_2 z = \frac{z}{z_4}. \end{aligned} \tag{2.35}$$

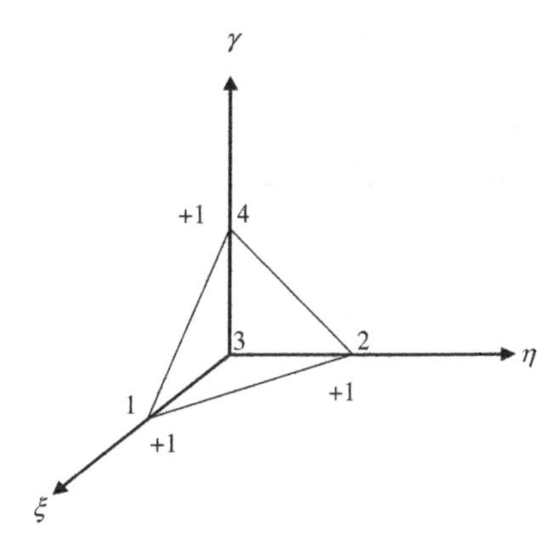

Fig. 2.6 Tetrahedral element in the local coordinate system (ξ, η, γ).

Again adopting a coordinate system of (ξ, η, γ) as shown in Fig. 2.6 we have

$$\begin{aligned} \xi &= \frac{x}{x_1}, \\ \eta &= \frac{y}{y_2}, \\ \gamma &= \frac{z}{z_4}, \end{aligned} \tag{2.36}$$

Replacing from Eq. (2.36) into Eq. (2.35) we get

$$N_1 = \xi, \quad N_2 = \eta, \quad N_3 = 1 - \xi - \eta - \gamma, \quad N_4 = \gamma. \tag{2.37}$$

2.3 Some Examples of Complex and Multiplex Elements

In this section we consider the derivation of the shape functions of a number of complex and multiplex elements. It is shown that in some cases a method similar to the technique based on the direct use of interpolation procedures can be employed. However, there are cases that an indirect approach must be adopted to derive the required elemental shape functions.

2.3.1 *Quadratic and cubic one-dimensional elements*

Consider the three-node one-dimensional element shown in Fig. 2.7. Within this element we can use a complete quadratic function to approximate a given field variable T as

$$\tilde{T} = ax^2 + bx + c. \tag{2.38}$$

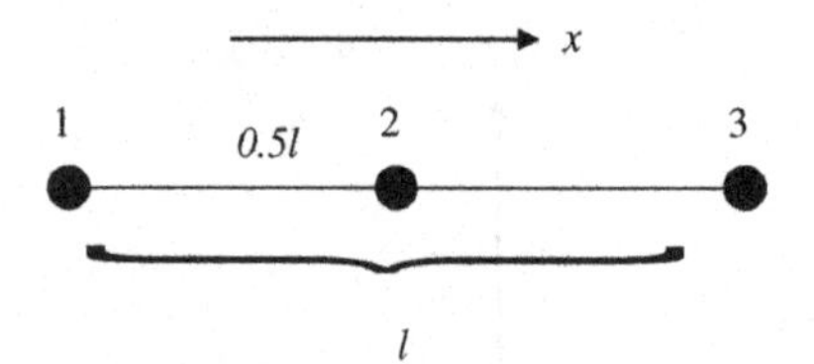

Fig. 2.7 One-dimensional quadratic element of length l. Coordinate system used is $x(0,l)$.

Therefore at the nodes of this element we have

$$\begin{cases} T_1 = ax_1^2 + bx_1 + c \\ T_2 = ax_2^2 + bx_2 + c \, , \\ T_3 = ax_3^2 + bx_3 + c \end{cases} \tag{2.39}$$

and using matrix notation

$$\begin{bmatrix} T_1 \\ T_2 \\ T_3 \end{bmatrix} = \begin{bmatrix} 1 & x_1 & x_1^2 \\ 1 & x_2 & x_2^2 \\ 1 & x_3 & x_3^2 \end{bmatrix} \begin{bmatrix} a \\ b \\ c \end{bmatrix} . \tag{2.40}$$

We can now derive the general form of the shape functions associated with this element as

$$N = PC^{-1} = \begin{bmatrix} 1 & x & x^2 \end{bmatrix} \begin{bmatrix} 1 & x_1 & x_1^2 \\ 1 & x_2 & x_2^2 \\ 1 & x_3 & x_3^2 \end{bmatrix}^{-1} \tag{2.41}$$

and

$$N = PC^{-1} = \begin{bmatrix} 1 & x & x^2 \end{bmatrix} \frac{1}{|C|} \begin{bmatrix} \begin{vmatrix} x_2 & x_2^2 \\ x_3 & x_3^2 \end{vmatrix} & -\begin{vmatrix} 1 & x_2^2 \\ 1 & x_3^2 \end{vmatrix} & \begin{vmatrix} 1 & x_2 \\ 1 & x_3 \end{vmatrix} \\ -\begin{vmatrix} x_1 & x_1^2 \\ x_3 & x_3^2 \end{vmatrix} & \begin{vmatrix} 1 & x_1^2 \\ 1 & x_3^2 \end{vmatrix} & -\begin{vmatrix} 1 & x_1 \\ 1 & x_3 \end{vmatrix} \\ \begin{vmatrix} x_1 & x_1^2 \\ x_2 & x_2^2 \end{vmatrix} & -\begin{vmatrix} 1 & x_1^2 \\ 1 & x_2^2 \end{vmatrix} & \begin{vmatrix} 1 & x_1 \\ 1 & x_2 \end{vmatrix} \end{bmatrix}^T . \tag{2.42}$$

Replacing for the coordinates $x_1 = 0,\ x_2 = 0.5l$, and $x_3 = l$ then

$$|C| = \frac{l^3}{4}. \tag{2.43}$$

Therefore

$$N = \frac{4}{l^3}\begin{bmatrix}1 & x & x^2\end{bmatrix}\begin{bmatrix}0.25l^3 & 0 & 0\\ -0.75l^2 & l^2 & -0.25l^2\\ 0.5l & -l & 0.5l\end{bmatrix} \tag{2.44}$$

and

$$N = \frac{4}{l^3}\left[0.25l^3 - 0.75l^2x + 0.5lx^2, l^2x - lx^2, -0.25l^2x + 0.5lx^2\right], \tag{2.45}$$

and finally

$$\begin{cases} N_1 = 1 - 3\dfrac{x}{l} + 2\dfrac{x^2}{l^2} \\ N_2 = 4\dfrac{x}{l} - 4\dfrac{x^2}{l^2} \\ N_3 = -\dfrac{x}{l} + 2\dfrac{x^2}{l^2} \end{cases}. \tag{2.46}$$

We now consider the use of a normalized coordinate system as shown in Fig. 2.8.

To define this element the shape functions derived in the original coordinates can be transformed to the normalized system using the transformation relationship of $x = \frac{l}{2}(1+\xi)$. Therefore,

$$\begin{cases} N_1 = -\dfrac{1}{2}\xi(1-\xi) \\ N_2 = (1-\xi^2) \\ N_3 = \dfrac{1}{2}\xi(1+\xi) \end{cases}. \tag{2.47}$$

Direct application of quadratic Lagrangian interpolation can therefore provide the required shape functions for the described complex element. Similarly, if we use a cubic Lagrangian interpolation model we can derive the shape functions associated with the four-node one-dimensional cubic complex element shown in Fig. 2.9 (Li, 2006).

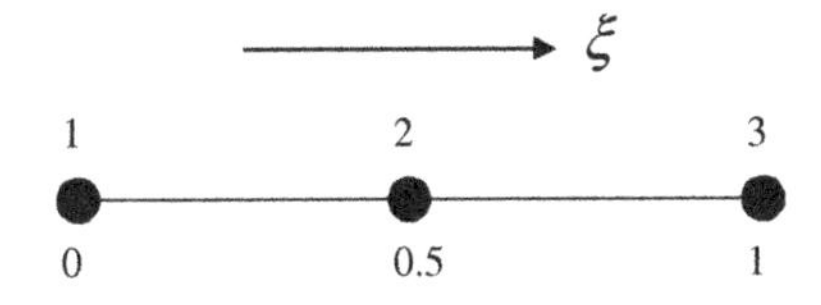

Fig. 2.8 One-dimensional quadratic element in the local coordinate system of $\xi(-1, +1)$.

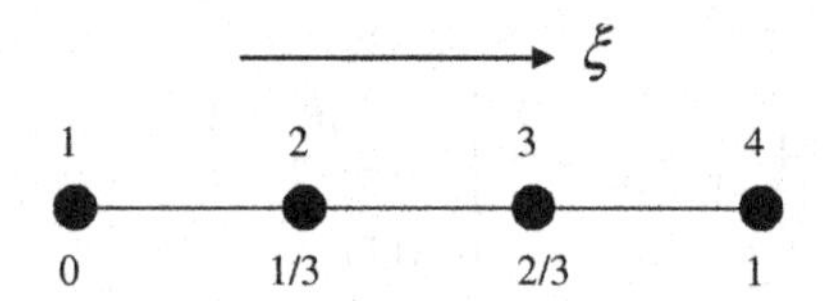

Fig. 2.9 One-dimensional cubic element coordinate system $\xi(-1,+1)$.

The shape functions associated with this element are found as

$$\begin{cases} N_1 = -\dfrac{9}{16}\left(\dfrac{1}{9}-\xi^2\right)(1-\xi) \\ N_2 = \dfrac{27}{16}(1-\xi^2)\left(\dfrac{1}{3}-\xi\right) \\ N_3 = \dfrac{27}{16}(1-\xi^2)\left(\dfrac{1}{3}+\xi\right) \\ N_4 = -\dfrac{9}{16}\left(\dfrac{1}{3}+\xi\right)(1-\xi^2) \end{cases}. \tag{2.48}$$

In two- and three-dimensional cases, shape functions associated with other members of the complex quadratic elements family, such as six-node triangle, can be derived via direct application of the two- or three-dimensional Lagrange interpolation polynomials.

2.3.2 *Four-node two-dimensional element*

The shape functions of multiplex rectangular and hexahedral elements can be derived by an indirect procedure using tensor product of one-dimensional shape functions. The interpolation polynomial for a bilinear rectangular element (see Fig. 2.10) is written as (Nassehi, 2002)

$$\tilde{T} = a + bx + cy + dxy. \tag{2.49}$$

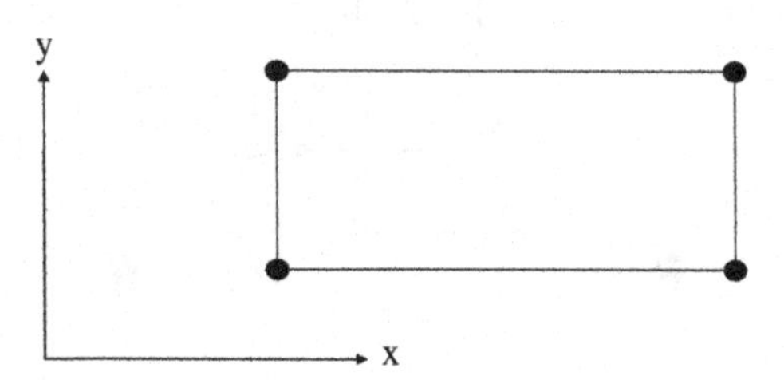

Fig. 2.10 Bilinear rectangular element.

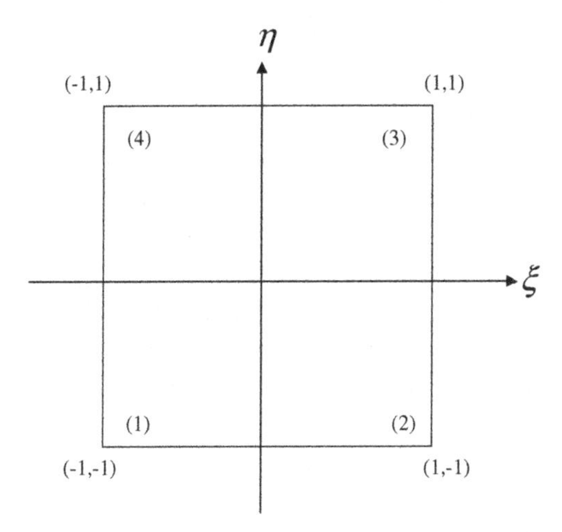

Fig. 2.11 Local coordinate system (ξ, η) associated with four-node multiplex element.

Note that the polynomial in Eq. (2.49) does not include all of the terms of a complete quadratic polynomial. However, it includes a second-order term and is not a linear two-dimensional function.

The right-hand side of Eq. (2.40) can be written as the product of two first-order polynomials in terms of x and y as

$$\tilde{T} = (a_1 + b_1 x)(a_2 + b_2 y). \tag{2.50}$$

Therefore to derive the shape function of a bilinear element the product of linear interpolation functions can be used. The four-noded rectangular element constructed in this way is called bilinear element. Using a normalized local coordinate system as shown in Fig. 2.11.

The shape functions of a bilinear element can be derived as

$$\begin{cases} N_1 = \frac{1}{2}(1-\xi)\frac{1}{2}(1-\eta) \\ N_2 = \frac{1}{2}(1+\xi)\frac{1}{2}(1-\eta) \\ N_3 = \frac{1}{2}(1+\xi)\frac{1}{2}(1+\eta) \\ N_4 = \frac{1}{2}(1-\xi)\frac{1}{2}(1+\eta) \end{cases}. \tag{2.51}$$

Similarly the shape functions of an eight-noded hexahedral element (eight-noded brick element) are found using tensor products of linear functions in x, y, and z directions. The interpolation polynomial for an eight-noded hexahedral element is written as

$$\tilde{T} = a + bx + cy + dz + exy + fxz + gzy + hxyz, \tag{2.52}$$

where a, b, c, d, e, f, g, and h are linearly independent coefficients that should be determined. The right-hand side of Eq. (2.52) can be written as the product of the polynomials in terms of x, y, and z

$$\tilde{T} = (a_1 + b_1 x)(a_2 + b_2 . y)(a_3 + b_3 z). \tag{2.53}$$

Again, using a normalized local coordinate system (Fig. 2.12) the shape functions of the trilinear element are written as

$$\begin{cases} N_1 = \frac{1}{2}(1-\xi)\frac{1}{2}(1-\eta)\frac{1}{2}(1-\gamma) \\ N_2 = \frac{1}{2}(1+\xi)\frac{1}{2}(1-\eta)\frac{1}{2}(1-\gamma) \\ N_3 = \frac{1}{2}(1+\xi)\frac{1}{2}(1+\eta)\frac{1}{2}(1-\gamma) \\ N_4 = \frac{1}{2}(1-\xi)\frac{1}{2}(1+\eta)\frac{1}{2}(1-\gamma) \\ N_5 = \frac{1}{2}(1-\xi)\frac{1}{2}(1-\eta)\frac{1}{2}(1+\gamma) \\ N_6 = \frac{1}{2}(1+\xi)\frac{1}{2}(1-\eta)\frac{1}{2}(1+\gamma) \\ N_7 = \frac{1}{2}(1+\xi)\frac{1}{2}(1+\eta)\frac{1}{2}(1+\gamma) \\ N_8 = \frac{1}{2}(1-\xi)\frac{1}{2}(1+\eta)\frac{1}{2}(1+\gamma) \end{cases} . \tag{2.54}$$

2.4 Convergence of Finite Element Approximations

Theoretically, the sequence of approximate solutions generated by a finite element scheme for a differential equation will converge to the exact solution after successive mesh refinement provided that the elemental interpolation polynomials satisfy the following convergence requirements (Rao, 1989).

- The field variable must remain continuous within the elemental spaces. This requirement is easily satisfied by choosing continuous functions as interpolation models.
- All of the uniform states of the field variable, up to the highest order appearing in the weighted residual functional, must be present in the polynomial at the limit when element size is reduced to zero.

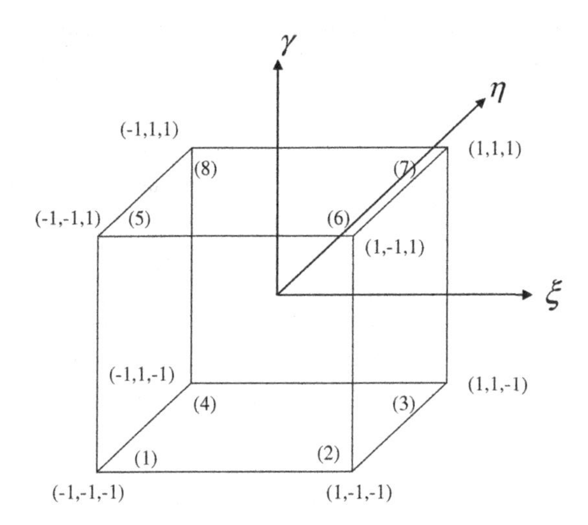

Fig. 2.12 Local coordinate system (ξ, η, γ) for hexahedral element (brick element).

- The field variable and its partial derivatives up to one order less than the highest order derivative appearing in the weighted residual functional must be continuous at element boundaries or interfaces.

In practice, however, infinite mesh refinement is not possible and some degree of approximation must be tolerated.

Within the described restriction in accuracy there are four basic approaches which can be used to improve a finite element approximation. These are:

- Reducing the element size (h-refinement)
- Using higher-order interpolation polynomials (p-refinement)
- Rearranging nodal points in a fixed element topology (r-refinement)
- Defining a new mesh with a better distribution of elements for a given problem.

Various combinations of these approaches are also possible.
In general, a smaller number of higher-order elements yields more accurate results compared to a larger number of simpler elements for the same overall effort. However, this does not mean that higher-order elements always generate better results. The main reason for this is that the savings resulting from the use of higher-order elements may become overshadowed by the

required increased effort in formulating and evaluating elemental stiffness matrices (Rao, 1989; Huebner *et al.*, 2001).

2.5 Continuity Conditions

In the finite element method, continuous field variables are represented in terms of "piecewise continuous" functions, defined over a single element. To carry out the integration of the derivatives of a field variable these stepwise functions should remain bounded in the interval of integration. Thus, to integrate the nth order derivative, the function must be continuous to the order of $(n-1)$ to make sure that any discontinuity in the nth derivative of the function remains finite. If nth derivative of the field variable (i.e. the highest order derivative) is continuous it is said to be C^n continuous within the element and C^{n-1} at element interfaces.

Derivation of elements that ensure C^0 continuity is straightforward and, in theory, the number of elements capable of satisfying C^0 continuity is infinite. This is because it is possible to add nodes and degrees of freedom to a given element to construct higher-order elements.

C^1 continuous elements can also be constructed using Hermite interpolation models (Nassehi, 2002), however, construction of elements which ensure C^2 and higher continuity is not straightforward and approximation techniques other than interpolation may be required to construct such elements.

2.6 Solved Examples

In Chap. 1, using the solution of benchmark problems as examples, the basic procedure of the weighted residual finite element model was illustrated. In this chapter, we present a number of solved problems to describe situations in which the basic procedure may not generate acceptable solutions or its effective use depends on using unrealistically refined meshes.

2.6.1 *Example 1: Limitation of the standard Galerkin procedure*

Consider the solution of the following equation

$$\begin{cases} \dfrac{d^2T}{dx^2} + aT = 10 \quad \text{in } \Omega \\ \text{subject to:} \\ T_A = 1, \quad T_B = 4 \end{cases}, \tag{2.55}$$

where Ω is a straight line AB of length 1 (Fig. 1.4, Chap. 1). After the discretization of AB into four linear elements of equal size the standard Galerkin procedure is implemented. Therefore, the elemental stiffness equation is derived as

$$\begin{bmatrix} -\int_{\Omega_e}\left(\frac{dw_1}{dx}\frac{dN_1}{dx} - aw_1N_1\right)dx & -\int_{\Omega_e}\left(\frac{dw_1}{dx}\frac{dN_2}{dx} - aw_1N_2\right)dx \\ -\int_{\Omega_e}\left(\frac{dw_2}{dx}\frac{dN_1}{dx} - aw_2N_1\right)dx & -\int_{\Omega_e}\left(\frac{dw_2}{dx}\frac{dN_1}{dx} - aw_2N_2\right)dx \end{bmatrix}\begin{Bmatrix} T_1 \\ T_2 \end{Bmatrix}$$
$$= \begin{Bmatrix} -w_1\Phi\,|_{\Gamma_e} + \int_{\Omega_e} 10w_1 dx \\ -w_2\Phi\,|_{\Gamma_e} + \int_{\Omega_e} 10w_2 dx \end{Bmatrix}, \tag{2.56}$$

where Φ represents the boundary line term.

We now compare three cases in which a takes the values of -1, -50, and -500, respectively.

For $a = -1$ the left-hand side of Eq. (2.56) gives

$$\begin{bmatrix} 4.0833 & -3.9583 \\ -3.9583 & 4.0833 \end{bmatrix}$$

and its corresponding global system of equations is

$$\begin{bmatrix} 8.1667 & -3.9583 & 0 \\ -3.9583 & 8.1667 & -3.9583 \\ 0 & -3.9583 & 8.1667 \end{bmatrix}\begin{Bmatrix} T_2 \\ T_3 \\ T_4 \end{Bmatrix} = \begin{Bmatrix} 1.4583 \\ -2.5 \\ 13.3333 \end{Bmatrix}. \tag{2.57}$$

The solution of this equation gives

$$\begin{cases} T_1 = 0.7013 \\ T_2 = 1.0785\;. \\ T_3 = 2.1554 \end{cases}$$

In Fig. 2.13 the comparison of these results with their corresponding analytical counterparts is shown.

For $a = -50$ the left-hand side of Eq. (2.56) gives

$$\begin{bmatrix} 8.1667 & -1.9167 \\ -1.9167 & 8.1667 \end{bmatrix}$$

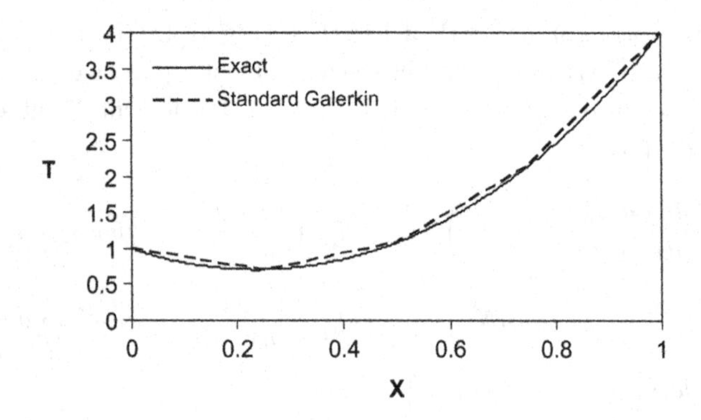

Fig. 2.13 $a = -1$, solution is stable and accurate. The error at $x = 0.5$ is 0.387%.

and the global system is found as

$$\begin{bmatrix} 16.3333 & -1.9167 & 0 \\ -1.9167 & 16.3333 & -1.9167 \\ 0 & -1.9167 & 16.3333 \end{bmatrix} \begin{Bmatrix} T_2 \\ T_3 \\ T_4 \end{Bmatrix} = \begin{Bmatrix} -0.5833 \\ -2.5 \\ 5.1667 \end{Bmatrix} \tag{2.58}$$

Therefore, the solution is

$$\begin{cases} T_2 = -0.0502 \\ T_3 = -0.1235 \, . \\ T_4 = 0.3018 \end{cases}$$

A comparison of this set of results with their analytical counterparts is shown in Fig. 2.14 and as shown in this graph, the standard Galerkin procedure based on the use of ordinary elements has produced somewhat

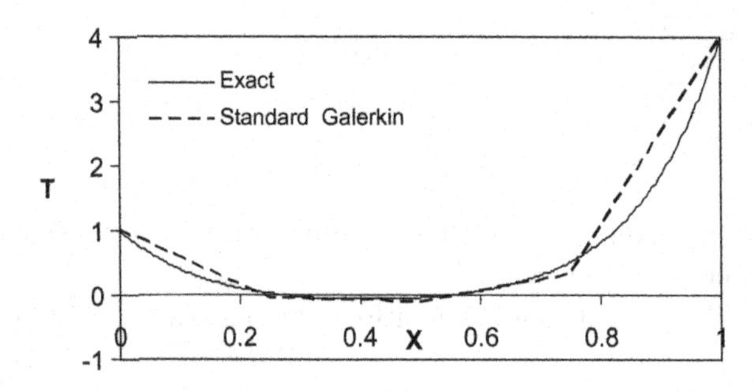

Fig. 2.14 Solution appears to be stable but it is inaccurate, the error at $x = 0.5$ is 1.67%.

inaccurate results. Therefore, in this case, in order to obtain a solution with a degree of accuracy comparable with the previous case further mesh refinement should be carried out.

We now consider a case in which $a = -500$. After the insertion of this value the left-hand side of Eq. (2.56) becomes

$$\begin{bmatrix} 45.6667 & 16.8333 \\ 16.8333 & 45.6667 \end{bmatrix}.$$

Therefore, the global system of finite element equations in this case is

$$\begin{bmatrix} 91.3333 & 16.8333 & 0 \\ 16.8333 & 91.3333 & 16.8333 \\ 0 & 16.8333 & 91.3333 \end{bmatrix} \begin{Bmatrix} T_2 \\ T_3 \\ T_4 \end{Bmatrix} = \begin{Bmatrix} -19.3333 \\ -2.5 \\ -69.8333 \end{Bmatrix} \tag{2.59}$$

Solution of Eq. (2.59) gives

$$\begin{cases} T_2 = -0.2418 \\ T_3 = 0.1637 \\ T_4 = -0.7948 \end{cases}.$$

A comparison of this result with the analytical solution in Fig. 2.15 shows that in this case the routine implementation of the standard Galerkin finite element procedure has generated inaccurate and unstable results.

Obviously we need a significantly more refined mesh to reduce the error in the numerical solution to a tolerable level. In more complex multidimensional problems, the degree of required mesh refinement may preclude the development of a practical solution scheme. Our basic proposition is, therefore, to develop methods that reduce or eliminate the need for excessive mesh refinement.

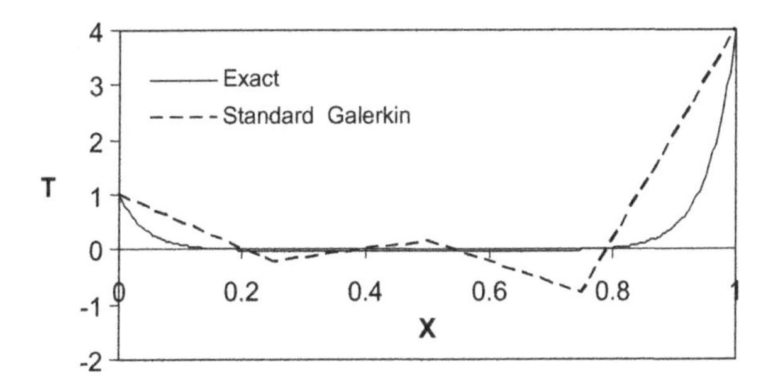

Fig. 2.15 Solution is unstable and inaccurate, the error at x = 0.5 is 4.82%.

2.6.2 *Example 2: A further problem associated with the use of the standard Galerkin finite element method*

We now consider the solution of a more complex differential equation which contains both first- and second-order terms

$$\begin{cases} \dfrac{d^2T}{dx^2} + 50\dfrac{dT}{dx} + T = 0 \quad \text{in } \Omega \\ \text{subject to:} \\ T_A = 0, \quad T_B = 1, \end{cases} \tag{2.60}$$

where Ω is a straight line AB of length 2. Equation (2.60) represents a transport problem in which the property in question is transported by both molecular and convection mechanisms. The coefficient of the first-order derivative in Eq. (2.60) is the ratio of convection velocity to the diffusivity and can, therefore, be treated as a function of the Peclet number (P_e) (Tucker, 1989). After the division of AB into 10 linear elements of equal size and application of the standard Galerkin procedure, we obtain the following elemental stiffness matrix

$$\begin{bmatrix} -29.934 & 30.033 \\ -19.967 & 20.066 \end{bmatrix}.$$

Assembly of the elemental equations over the common nodes and imposition of the boundary conditions results in the following global system of equations

$$\begin{bmatrix} -9.868 & 30.033 & 0 & 0 & 0 & 0 & 0 & 0 & 0 \\ -19.967 & -9.868 & 30.033 & 0 & 0 & 0 & 0 & 0 & 0 \\ 0 & -19.967 & -9.868 & 30.033 & 0 & 0 & 0 & 0 & 0 \\ 0 & 0 & -19.967 & -9.868 & 30.033 & 0 & 0 & 0 & 0 \\ 0 & 0 & 0 & -19.967 & -9.868 & 30.033 & 0 & 0 & 0 \\ 0 & 0 & 0 & 0 & -19.967 & -9.868 & 30.033 & 0 & 0 \\ 0 & 0 & 0 & 0 & 0 & -19.967 & -9.868 & 30.033 & 0 \\ 0 & 0 & 0 & 0 & 0 & 0 & -19.967 & -9.868 & 30.033 \\ 0 & 0 & 0 & 0 & 0 & 0 & 0 & -19.967 & -9.868 \end{bmatrix}$$

$$\times \begin{bmatrix} T_2 \\ T_3 \\ T_4 \\ T_5 \\ T_6 \\ T_7 \\ T_8 \\ T_9 \\ T_{10} \end{bmatrix} = \begin{bmatrix} 0 \\ 0 \\ 0 \\ 0 \\ 0 \\ 0 \\ 0 \\ 0 \\ -30.033 \end{bmatrix}.$$

However, the solution of the above given system of global equations generates an oscillatory and useless result (Fig. 2.16).

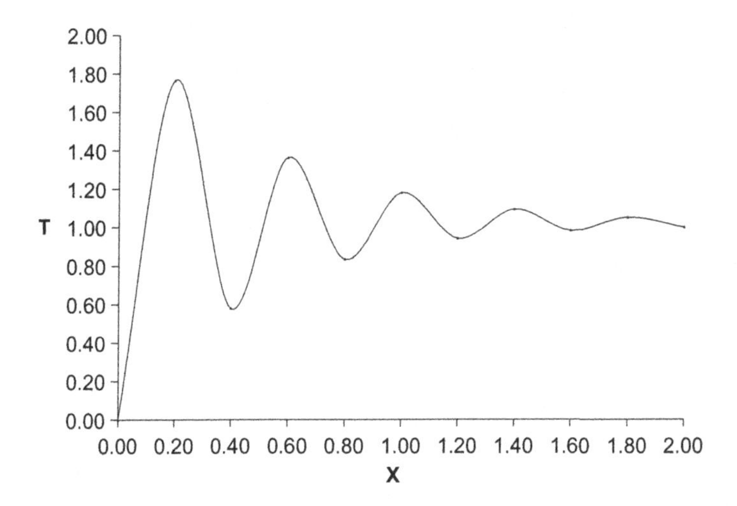

Fig. 2.16 Unstable result obtained using the standard Galerkin method.

The usual way of dealing with this type of problem is the use of the Petrov–Galerkin schemes (Chap. 1). In particular, Streamline Upwind Petrov–Galerkin (SUPG) method is the most commonly used technique in this type of problem, normally described as convection dominated transport problems. In the SUPG technique, instead of a weight function which is identical to the interpolation (shape) functions, a modified function based on the following weight function, originally described by Brooks and Hughes (1982), is used to construct the required weighted residual statement.

$$W_I = N_I + \gamma\, N_{I,i} \frac{\bar{v}_i}{|v|^2}, \tag{2.61}$$

where N_I and $N_{I,i}$ represent a shape function and its derivative, v is the convection velocity, γ is called an upwinding parameter, and summation convention over i is assumed. In a one-dimensional problem such as the one discussed here, the upwinded weight function can be written as $W_i = N_i + \beta \frac{dN_i}{dx}$. Therefore, dividing the problem domain into 10 linear elements of equal size and after the application of the described "upwinded weight" function, the elemental stiffness equation for the present example is found as

$$\begin{bmatrix} (-29.933 + 250.0\beta) & (30.033 - 250.0\beta) \\ (-19.967 - 250.0\beta) & (20.067 + 250.0\beta) \end{bmatrix} \begin{pmatrix} T_I \\ T_{II} \end{pmatrix} = \begin{pmatrix} q_I \\ -q_{II} \end{pmatrix}. \tag{2.62}$$

After the assembly of elemental equations into a global set and imposition of the boundary conditions, we can find the final solution of the original differential equation. However, the described solution depends on choosing an appropriate value for the upwinding parameter β. Depending on the value of this parameter, the final solution of the convection dominated problem can be oscillatory (unstable), or it can become stable but over-damped and inaccurate, and only for a specific value of this coefficient can a stable and accurate solution be generated (Nassehi, 2002). For one-dimensional problems this value is calculated using the following relationship (Hughes and Brooks, 1979)

$$\gamma = \frac{vh}{2}\left[\coth\left(\frac{P_e}{2}\right) - \frac{2}{P_e}\right], \tag{2.63}$$

where h is the length of individual linear elements in the computational mesh and $P_e = \frac{vh}{\alpha}$ is the mesh Peclet number which in this case is equal to 10. Using this value the coefficient γ is found to be 4.0 and hence the optimum value for β in this problem is -0.08. Insertion of $\beta = -0.08$ into Eq. (2.62) gives the elemental form of the stiffness equation for this problem. The analytical solution of Eq. (2.60) is

$$T = 1.0408(e^{-0.02x} - e^{-49.98x}). \tag{2.64}$$

In Fig. 2.17 this solution is compared with the result obtained using upwinded finite element scheme. As this figure shows, the streamlined upwind finite element procedure has generated a very accurate result for this problem. However, the application of this technique in multidimensional problems is not straightforward. This is mainly because that theoretical extension of Eq. (2.63) to obtain an optimal value for the upwinding parameter is not possible. Over the past decades, therefore, various investigators

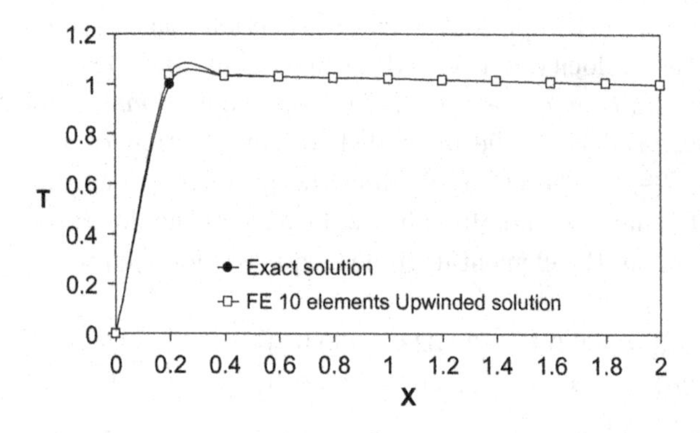

Fig. 2.17 Comparison of analytical and streamlined upwind solutions.

have suggested heuristic ways of finding the required upwinding parameter under various conditions. For example, in two-dimensional problems the following relationship is sometimes used (Petera *et al.*, 1993)

$$W_I = N_I + \gamma_c \frac{|h_1 v_1 + h_2 v_2|}{2\,|\bar{\mathbf{v}}|^2} \left(v_1 \frac{\partial N_I}{\partial x_1} + v_2 \frac{\partial N_I}{\partial x_2} \right). \tag{2.65}$$

In general, however, due to the impossibility of the selection of an optimum value for γ_c that can guarantee the elimination of all spurious cross-wind diffusion (Hughes and Brooks, 1979), and, indeed, the uncertainty inherent in the definition of element length in multidimensional domains, SUPG-based solutions involve some degree of upwinding error.

2.6.3 *Bubble function enriched elements*

We now introduce the concept of enriched elements which can be used in conjunction with the standard Galerkin method to overcome the difficulties explained in the previous two examples.

Consider the linear element shown in Fig. 2.18. Ignoring the fact that there are three nodes in this element we take the interpolation functions associated with nodes 1 and 2, (corresponding, respectively, to degrees of freedom T_1 and T_2), as ordinary linear Lagrangian shape functions. However, we define a special shape function corresponding to the middle node (i.e. relevant to the degree of freedom T_b) as $N_b = x(l - x)$.

Therefore, the approximation for the unknown field variable in one-dimensional problems, such as those discussed in the previous sections, is now written as

$$\tilde{T} = \left(1 - \frac{x}{l}\right) T_1 + \frac{x}{l} T_2 + x\,(l - x)\, T_b. \tag{2.66}$$

The important point to note is that the introduction of the additional degree of freedom is not going to affect the basic structure of the solution procedure or the degree of the inter-element continuity. This is because we have deliberately chosen a shape function for the middle node which disappears on all of the exterior boundaries of the element. This type of function is called a "bubble function", indicating that it will generate a

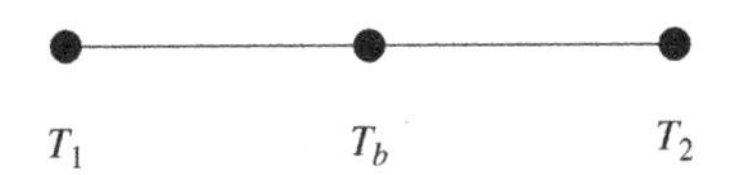

Fig. 2.18 An enriched finite element.

nonzero value only in the interior of the elements. The enrichment of the approximation of the unknown function by the addition of the "bubble function" $\phi_b = x\,(l - x)$ to the normal Lagrangian interpolation is, therefore, restricted to the interior of the element. This results in a significant increase in the accuracy of the approximation of the field unknown over an element domain without altering the basic properties of the finite element scheme. At this point it appears we have chosen this function arbitrarily. However, in the following chapters of this book detailed procedures for the derivation of "bubble functions" are described. After the introduction of the approximation defined by Eq. (2.66) for the field unknown, we continue with the standard Galerkin finite element solution of any given equation in the usual manner. However, as is explained in the following chapters of this book, specific manipulations are required to render the final set of equations obtained in such a scheme determinate. As an example of the power of bubble function enriched elements, Fig. 2.19 shows the comparison between the analytical solution and the Galerkin finite element solution obtained for Eq. (2.60) utilizing this type of element.

In Fig. 2.19 dots used to mark the position of nodal analytical solutions are all covered by the finite element results, clearly demonstrating the superiority of the bubble elements scheme over the used streamlined upwinding scheme. In Chap. 3 more detailed explanations about both upwinding and bubble function enriched techniques are given and techniques that can be used to derive appropriate shape functions of bubble function enriched elements are described.

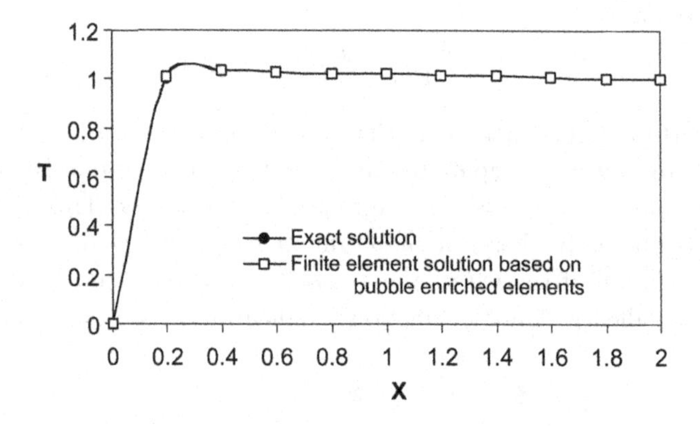

Fig. 2.19 Comparison of the solutions obtained using analytical and enriched element methods.

References

[1] Brooks, A.N. and Hughes T.J.R., Streamline Upwind/Petrov–Galerkin formulations for convection dominated flows with particular emphasis on the incompressible Navier–Stokes equations, *Comput. Meth. Appl. Mech. Eng.*, 1982; 32; 199–259.

[2] Huebner, K.H., Dewhirst, D.L., Smith, D.E. and Byrom, T.G., *The Finite Element Method for Engineering*, John Wiley & Sons Inc., 2001.

[3] Hughes, T.J.R. and Brooks, A.N., A multidimensional upwind scheme with no crosswind diffusion, in *Finite Element Methods for Convection Dominated Flows*, AMD; T.J.R. Hughes (Ed.), ASME, New York, 1979; Vol. 34.

[4] Li, B.Q., *Discontinuous Finite Elements in Fluid Dynamics and Heat Transfer*, Springer-Verlag, London, 2006.

[5] Nassehi, V., *Practical Aspects of Finite Element Modelling of Polymer Processing*, John Wiley & Sons, Chichester, 2002.

[6] Petera, J., Nassehi, V. and Pittman, J.F.T., Petrov–Galerkin methods on isoparametric bilinear and biquadratic elements tested for a scalar convection–diffusion problem, *Int. J. Numer. Meth. Heat Fluid Flow*, 1993; 3; 205–221.

[7] Rao, S.S., *The Finite Element Method in Engineering*, Pergamon Press, 1989.

[8] Tucker, C.L. III (Ed.). *Computer Modeling for Polymer Processing*, Hanser Publishers, Munich, 1989.

CHAPTER 3

Basic Concepts of Multiscale Finite Element Modeling

In this chapter, using worked examples, principle concepts of a number of multiscale finite element models are introduced. The chapter covers multiscale variational formulation, residual free bubble (RFB) function technique, and its application to practical problems and direct static condensation (STC) methods. To maintain simplicity the focus in this chapter is on the construction of schemes for one-dimensional problems. Here the solution of one-dimensional diffusion–reaction (DR), convection–diffusion (CD), and the general convection–diffusion–reaction (CDR) equations are considered as the benchmark cases.

In general the solution of multiscale problems with standard finite element schemes gives rise to unstable and oscillatory results. Over the past decades various techniques have been developed which can stabilize finite element solutions. In this chapter, we start with a brief study of the most common stabilization techniques that are used in finite element schemes.

3.1 Stabilization of Finite Element Schemes

In Chap. 2, an introductory example in which an upwinding procedure is used to stabilize the finite element solution has been presented. In this section, we expand this subject and examine the conditions that require the application of stabilization techniques. In line with historical development of the upwinding methods we focus on the solution of convection dominated transport processes to explain the fundamental concepts of the stabilization of finite element schemes.

Transport phenomena such as heat conduction and mass diffusion, similar to many other types of problems encountered in solid mechanics, are expressed in terms of differential equations in which the differential operator is symmetric (the simplest form of a self-adjoint operator). In this case, the application of the standard Galerkin technique in conjunction with finite element discretizations results in the construction of symmetric stiffness equations and, in general, yield accurate and stable solutions. In contrast, whenever the transport mechanism involves a significant degree of convection the symmetry of the differential operator in the governing equation of the problem is lost and the application of the Galerkin method results in a unsymmetric stiffness matrix. This can lead to an oscillatory and unstable solution. Theoretically, this problem can be resolved by refining the computational mesh to a degree that convection can no longer be regarded as the dominant transport mechanism within the elements of the mesh (Hughes and Brooks, 1979). However, this method is not computationally cost effective and alternative techniques should be employed to stabilize convection dominated transport problems.

Initially the analogy between upwind finite difference schemes for convection dominated problems was used as a guide by the investigators to develop stable finite element schemes. An example for this approach is the scheme developed by Christie *et al.* (1976) which was based on the modification of weighting functions for one-dimensional CD problems. This scheme was later generalized by Heinrich *et al.* (1977) to solve two-dimensional problems. These schemes, however, were prone to overdamp the solutions and produced unrealistic results. Brooks and Hughes (1982) proposed a method called streamline Petrov–Galerkin method which is more robust and significantly reduces the risks of artificial diffusion and overdamping of the results. A brief account of these approaches is outlined in the next section.

3.1.1 *Upwinded finite element schemes*

Consider the following scalar steady state convection–diffusion (CD) equation

$$-k\Delta T + a\nabla T = 0, \tag{3.1}$$

where T is the field unknown, k and a are coefficients and ∇ and Δ are gradient and Laplacian operators, respectively. The solution of Eq. (3.1) via the standard Galerkin method results in under-diffuse or unstable results depending on the relative dominance of the first-order (i.e. the convection) term. Such a solution can be stabilized by the modification of the weight function applied to the first-order term to produce an upwinding effect. This is analogous to the use of central difference for the second order and backward or forward difference for the first-order terms, and always generates over-diffuse (overdamped) results (Hughes and Brooks, 1979). Upwinding in a finite element context can be achieved via the application of one of the following:

- Artificial diffusion — in which the coefficient of the diffusion (second-order) term in Eq. (3.1) is modified by the addition of some artificial diffusion to the physical diffusion.
- Quadrature — in which the numerical integration of the convection term is modified to generate an upwinding effect (Hughes, 1978).
- Petrov–Galerkin method — in which the weighting used upwind of a node is different from the one applied in the downwind of that node. In Fig. 3.1 a schematic presentation of weight functions for the standard Galerkin and Petrov–Galerkin methods are shown.

Use of these techniques involves difficulties which can be resolved by the streamlining of the upwinding in the direction of flow (i.e. the direction of dominant convection).

3.1.2 *Upwinding techniques*

As shown in Chap. 1, the use of streamline upwind technique in one-dimensional cases generates very accurate results assisted by the availability of a relationship between an optimal value for the upwinding parameter and physical coefficients in the CD equation. However, as mentioned before, the extension of such a relationship to multidimensional situations is not possible. Another major problem with the simple streamline upwind scheme is that although its use removes the basic problem of cross-wind diffusion, it cannot adequately cope with source or sink and transient conditions. The result is the generation of over-diffusive solutions when such terms are present in the governing equations of a problem. Consistent application of the modified weight function to all terms in the CD equation resolves this difficulty. This method, known as the consistent streamline upwind

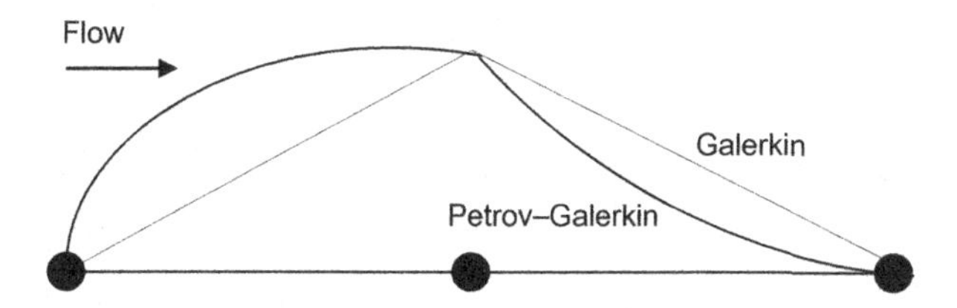

Fig. 3.1 Comparison of symmetric and modified weight functions used in the standard and Petrov–Galerkin methods (Brooks and Hughes, 1982).

Petrov–Galerkin (SUPG) method (Brooks and Hughes, 1980), is explained in this section.

In the standard Galerkin method the weighting functions are considered to be continuous across inter-element boundaries. The SUPG method uses discontinuous weighting functions (Brooks and Hughes, 1982) as

$$w^d = w + p, \tag{3.2}$$

where w is a continuous weight function and p is a discontinuous streamline upwind contribution added to it. Both w and p are assumed to be smooth within the domain of an element (Fig. 3.2).

Equation (3.2) is generally written as

$$w^d = w + \tau a \nabla w, \tag{3.3}$$

where the coefficient τ is represented (Codina, 1998) as

$$\tau = \frac{\gamma l}{2a}, \tag{3.4}$$

where

$$\gamma = \coth(P_e) - \frac{1}{P_e} \quad \text{and} \quad P_e = \frac{al}{2k}, \tag{3.5}$$

where l is the "element length."

The described modification of the weight function in the upwinding schemes can be generalized as follows. Consider the CDR operator written as

$$-k\Delta + a\nabla + s$$

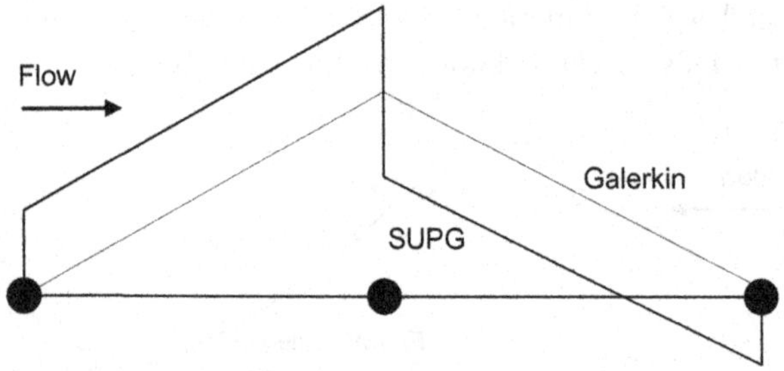

Fig. 3.2 Schematic representation of weight functions in the SUPG and standard Galerkin methods (Brooks and Hughes, 1982).

in which $k > 0$ is the diffusion coefficient, a is the convection velocity and $s \geq 0$ represents a source. We now define the terms of this operator as

$$\begin{cases} \bar{k} = k + \tau a^2 \\ \bar{a} = a - (\varsigma + 1)\tau a\, s\, . \\ \bar{s} = s - \varsigma \tau s^2 \end{cases} \tag{3.6}$$

The SUPG method corresponds to $\varsigma = 0$. Other stabilization techniques can also be derived using this general definition. The most effective amongst these techniques are the Galerkin Least Squares method (GLS) corresponding to $\varsigma = -1$ and the Subgrid method (SGS) corresponding to $\varsigma = 1$. The remaining parameter in Eq. (3.6) is derived as (Codina, 1998)

$$\tau = \frac{1}{\frac{4k}{l_e^2} + \frac{2a}{l_e} + s}. \tag{3.7}$$

Hauke (2002) proposed the following modification for the above statement to include cases where $s < 0$:

$$\tau = \frac{1}{\frac{4k}{l_e^2} + \frac{2|a|}{l_e} + |s|}. \tag{3.8}$$

These methods can also be used to solve time dependent nonhomogeneous problems.

3.2 Multiscale Approach

In general, accurate numerical modeling of a multiscale phenomenon depends on the level of discretization (i.e. mesh refinement and type of elemental approximation) used. Therefore, when using a coarse mesh any given solution scheme which does not take into account multiscale nature of a problem may generate unstable results. Therefore, to avoid the problems associated with the use of coarse meshes, as well as limiting computational costs, intrinsically stable finite element schemes techniques which can handle multiscale behavior need to be developed. To achieve this goal different techniques based on a multiscale variational method, RFB function method, and multilevel and discontinuous enrichment methods have been developed. In the following sections multiscale variational and RFB function methods are explained in detail. Multilevel and discontinuous enrichment methods are not discussed in this book.

3.2.1 *Variational multiscale method*

The development of variational multiscale and bubble function methods can be traced back to the construction of stabilized schemes. Both approaches can give rise to very effective schemes for dealing with difficulties associated with multiscale bahavior. However, the variational multiscale method provides a more general framework for the analysis of the stabilization technique and is discussed first here.

The basic concept of the variational multiscale method, first developed by Hughes (1995), is that the finite element schemes based on low-order polynomials as approximating functions do not yield robust solutions for problems involving fine scale features that are numerically irresolvable within the "length scale" of elements composing a computational mesh. In the variational multiscale method this difficulty is resolved by using separate approximations for different scale phenomena in a given problem (Juanes and Patzek, 2005). Therefore, a multiscale solution scheme consists of different solution strategies for different scales. The most natural decomposition of a solution scheme into different forms is to state a multiscale field variable as the sum of its fine and coarse scale components as $T = T_c + T_b$. This allows the use of a different solution strategy for each component, thereby eliminating the fine scale part of the unknown from the procedure used to find the coarse scale variations. The coarse part (the numerically resolvable part shown here as T_c) can be approximated by simple polynomial approximating functions in the usual manner, and the fine part, (numerically irresolvable or subgrid part shown here as T_b) can be derived analytically. Furthermore, the subgrid scale is made to vanish on the element boundaries and is basically represented in terms of elemental Green's function in conjunction with homogeneous boundary conditions. To illustrate these points, we consider the solution of the following problem:

$$\begin{cases} LT = f & \text{in } \Omega \\ T = c & \text{on } \Gamma \end{cases}, \tag{3.9}$$

where L is a linear differential operator, T is the unknown, and Ω is the problem domain with a smooth boundary defined as Γ. Equation (3.9) is used to construct a variational formulation as

$$a(T, w) = (LT, w), \tag{3.10}$$

where $a(.,.)$ is a bilinear form (Hughes, 1995), $(.,.)$ indicates a scalar product and w is an appropriate test (weight) function. Therefore,

$$a(T, w) = (f, w). \tag{3.11}$$

We now consider the decomposition of the field unknown and the weight function into resolvable and irresolvable parts (i.e. coarse and fine scales) as

$$T = T_c + T_b, \tag{3.12}$$

$$w = w_c + w_b, \tag{3.13}$$

where $w_b = T_b = 0$ on Γ_e, therefore,

$$a(T_c + T_b, w_c + w_b) = (f, w_c + w_b). \tag{3.14}$$

The variational formulation has now given rise to two sub-problems. The first problem, shown in terms of Eq. (3.15), can be used to solve the fine scale part in terms of the coarse scale variables.

$$a(T_c, w_b) + a(T_b, w_b) = (f, w_b). \tag{3.15}$$

This result is subsequently substituted into the second sub-problem (represented by Eq. 3.16) to obtain a modified form which only involves the coarse scale part of the field variable. Thus,

$$a(T_c, w_c) + a(T_b, w_c) = (f, w_c). \tag{3.16}$$

This procedure is called the STC technique.

The variational formulation of Eq. (3.15) can now be represented using the differential operator as

$$\begin{cases} LT_b = -(LT_c - f) & \text{in } \Omega_e \\ T_b = 0 & \text{on } \Gamma_e \end{cases}, \tag{3.17}$$

where $-(LT_c - f)$ is the residual generated by the approximation used for the resolved scale. Figure 3.3 gives a schematic representation of elemental domain Ω_e and elemental boundary Γ_e used in Eq. (3.17).

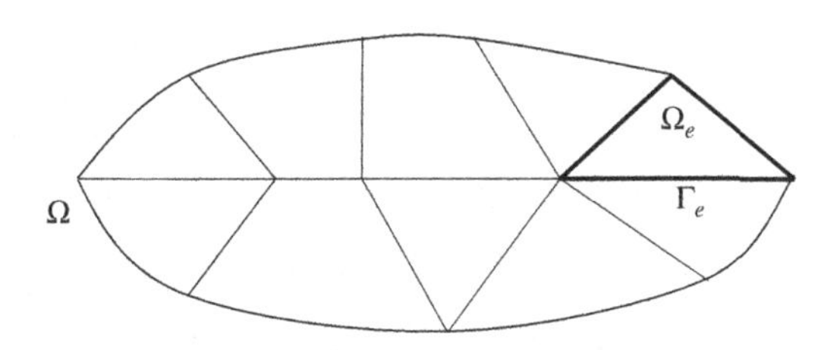

Fig. 3.3 Global and elemental domains and boundaries.

Solution of Eq. (3.17) leads to the subgrid scale model. Hughes (1995) has proposed to use Green's function technique to obtain such a solution. Green's function approach corresponding to Eq. (3.17) is written as

$$\begin{cases} L^* g = \delta & \text{in } \Omega_e \\ g = 0 & \text{on } \Gamma_e \end{cases}, \tag{3.18}$$

where L^* is an adjoint operator (see Appendix A.6). Thus,

$$T_b = -\int_{\Omega_e} g(LT_c - f) d\Omega_e \tag{3.19}$$

provides a solution for Eq. (3.17). The upwinding parameter τ in the SUPG method can also be related to the elemental Green's function as

$$\tau = \frac{1}{l_e} \int_{\Omega_e} g \, d\Omega_e \tag{3.20}$$

in which l_e is a characteristic element length.

Elimination of subgrid scales from the variational problem posed for the resolvable (i.e. coarse) scales in the described manner leads to a method capable of representing the effect of the fine scale variations. Note that the subgrid scale can be assumed to be zero at elemental boundaries. Further explanation of this concept is given in the next section via the introduction of bubble function-enriched elements.

3.2.2 *(RFB) function method*

Bubble functions are typically high-order polynomials which represent variations inside an element but vanish at the element boundaries. RFB function method is based on the solution of Eq. (3.17) in a way that the residual $-(LT_c - f)$ becomes zero within each element. This is achieved by deriving differential equations solvable within elements in terms of the subgrid scale functions using Eq. (3.17). RFB functions are derived via the analytical solution of these equations in conjunction with homogeneous boundary conditions, and should therefore strongly satisfy them within each element. Bubble functions and elemental Green's functions are related by Eq. (3.19). Therefore, previously described stabilization and RFB function techniques, and use of elemental Green's function are all interrelated procedures.

As mentioned before, the derivation of the bubble functions for a given problem is based on the analytical solution of its governing differential equation posed within each element (subject to homogeneous boundary conditions).

Let us consider a boundary value problem defined in $\Omega \subset R^2$ as (Franca and Russo, 1997a,b; Brezzi *et al.*, 1998)

$$\begin{cases} LT = f & \text{in } \Omega \\ T = 0 & \text{on } \Gamma \end{cases}, \tag{3.21}$$

where L is a linear differential operator and f is a given source function defined on the problem domain Ω. The standard Galerkin weighted residual statement of this problem is

$$a(T,v) = (LT,v) = (f,v), \tag{3.22}$$

where $a(.,.)$ is a bilinear form and $(.,.)$ represents a scalar product. In a problem involving both fine and coarse scale variations the unknown is decomposed to

$$T = T_c + T_b, \tag{3.23}$$

where T_b is the fine scale. Within the space of a finite element the coarse scale variations of the field unknown are approximated using normal interpolation (shape) functions. As mentioned earlier bubble functions are set to vanish on element boundaries and hence equations which correspond to these functions can be removed from the set of elemental equations via STC procedure (Donea and Huerta, 2003). To implement the STC we set $v = v_b$ in Eq. (3.22) to obtain

$$a(T_c + T_b, v_b)_{\Omega_e} = (f, v_b)_{\Omega_e}. \tag{3.24}$$

The subscript Ω_e indicates that formulation is restricted to individual elemental domains. However, as it is assumed that the bubble functions satisfy the differential equation in each element, that is to say,

$$LT_b = -(LT_c - f) \quad \text{in } \Omega_e, \tag{3.25}$$

the STC is automatically satisfied (Brezzi *et al.*, 1998).

Equation (3.25) is solved considering that $T_b = T_b^0 + T_b^f$, where, T_b^0 and T_b^f are, respectively, solutions of the following equations (Brezzi *et al.*, 1998)

$$\begin{cases} LT_b^0 = -LT_c & \text{in } \Omega_e \\ T_b^0 = 0 & \text{on } \Gamma_e \end{cases} \tag{3.26}$$

and

$$\begin{cases} LT_b^f = f & \text{in } \Omega_e \\ T_b^f = 0 & \text{on } \Gamma_e \end{cases}. \tag{3.27}$$

Assuming ϕ and ψ to be, respectively, bubble- and polynomial-based shape functions Eqs. (3.26) and (3.27) can be rewritten as

$$\begin{cases} L\phi_i = -L\psi_i & \text{in } \Omega_e \\ \phi_i = 0 & \text{on } \Gamma_e \end{cases} \tag{3.28}$$

and

$$\begin{cases} L\phi_f = f & \text{in } \Omega_e \\ \phi_f = 0 & \text{on } \Gamma_e \end{cases}, \tag{3.29}$$

where ψ_i and ϕ_i are functions associated with node i. Ω_e is the element domain and Γ_e is the element boundary. Hence

$$\tilde{T} = \sum_{i=1}^{n} T_i(\psi_i + \phi_i) + \phi_f, \tag{3.30}$$

where n is the number of nodes per element.

3.3 Practical Implementation of the RFB Function Method

The theoretical outline given in the previous sections provides an insight into the main concepts of the multiscale method described in this book. However, such analysis does not automatically lead to the construction of practical solution schemes. For example, identification of elemental Green's functions in most cases is not straightforward and can present significant complications. Furthermore, in many cases the multiscale nature of a physical problem may not be obvious and the difficulties detected during their solution should be used to evaluate the suitability of the employed method. Therefore, in this section steps that should be taken to construct unambiguous solution procedures for a range of multiscale problems are explained.

Note that elements in a computational mesh of a physical problem have, normally, different shapes and sizes. Therefore, the use of a global coordinate system results in a nonuniform range of characteristic element length and coordinate values. To avoid the complications resulting from this, we use a local coordinate system $x(0, l)$ in conjunction with isoparametric (mapped) elements in the following explanations. The elemental system is further normalized and, for example, in one-dimensional cases is represented as $\xi(-1,+1)$.

As mentioned before, construction of the RFB function method depends on the analytical solution of differential equations. This presents a severe restriction in cases where we need to handle partial differential equations (PDEs), as in general analytical solution of PDEs cannot be taken for granted. To overcome this complication we use a semi-discrete method to obtain solutions for PDEs (Parvazinia *et al.*, 2006b). Therefore the main steps of the procedure adopted here are summarized as:

Step 1. One-dimensional governing equation is presented within a linear Lagrangian element and solved over each node subject to the following elemental boundary conditions.

For node 1: $T = 1$ at $x = 0$ and $T = 0$ at $x = 1$.
For node 2: $T = 0$ at $x = 0$ and $T = 1$ at $x = 1$.

This solution provides an analytical expression for the functions used as shape functions in subsequent approximations.

Step 2. Using Taylor series expansion, the analytical solution found in step 1 is expanded upon. In practice only the first few terms of this expansion are kept and the rest are truncated.

Step 3. Terms of the truncated series are cast into forms with parts that vanish at element boundaries.

Step 4. After the derivation of the bubble-enriched shape functions, the problem is solved using the usual Galerkin finite element procedures.

To illustrate the above described process we consider the following example. Note that in all of the examples in this chapter, DR, CD and CDR equations are considered in dimensionless forms in a linear domain of unit length and solved subject to the following boundary conditions:

$$\begin{cases} T^*(0) = 0 \\ T^*(1) = 1 \end{cases}.$$

3.3.1 *Multiscale finite element solution of the DR equation using RFB function method*

Steady state DR equation in domain $\Omega \subset R^d$ can be written as follows

$$k\Delta T - sT = f, \tag{3.31}$$

where k is the diffusion coefficient and s is a source/sink term ($s < 0$ represents production and $s > 0$ stands for dissipation), T is the field unknown and f is a given source term. Using the following dimensionless forms

$$\begin{cases} T^* = \dfrac{T}{T_0} \\ \bar{x}^* = \dfrac{\bar{x}}{h} \end{cases}, \tag{3.32}$$

where T_0 is a reference value for the field variable, h is a characteristic length (e.g. width of the domain) and $\bar{x}$ represents position vector in the selected coordinate system. After substitution from Eq. (3.32), Eq. (3.31) is written in a dimensionless form as

$$\Delta T^* - D_a T^* = f^*, \tag{3.33}$$

where D_a is the Damköhler number and f^* is the dimensionless source term defined, respectively, as

$$\begin{cases} D_a = \dfrac{sh^2}{k} \\ f^* = \dfrac{h^2}{kT_0} f \end{cases}. \tag{3.34}$$

We now consider the implementation of the above described steps.

Step 1.

Solution of one-dimensional equation in conjunction with elemental boundary conditions. In its one-dimensional form Eq. (3.33) is written as

$$\frac{d^2T^*}{dx^{*2}} - D_a T^* = f^*. \tag{3.35}$$

Using two-node linear elements, the following nodal equations are derived from Eq. (3.35) for the shape functions N_1 and N_2

$$\begin{cases} \dfrac{d^2 N_1}{dx^2} - D_a N_1 = 0 \quad \text{for } x \in [0 - l] \\ \begin{cases} N_1(0) = 1 \\ N_1(l) = 0 \end{cases} \end{cases} \tag{3.36}$$

and

$$\begin{cases} \dfrac{d^2 N_2}{dx^2} - D_a N_2 = 0 \quad \text{for } x \in [0 - l] \\ \begin{cases} N_2(0) = 0 \\ N_2(l) = 1 \end{cases} \end{cases}, \tag{3.37}$$

where l is, in general, a characteristic element length. The solution of the above equation gives the analytical shape functions expressed in a local elemental coordinate system as

$$\begin{cases} N_1 = \dfrac{\sinh \sqrt{D_a}(l-x)}{\sinh \sqrt{D_a} l} \\ N_2 = \dfrac{\sinh \sqrt{D_a} x}{\sinh \sqrt{D_a} l} \end{cases}. \tag{3.38}$$

Step 2.

Hyperbolic functions (3.38) can be used directly if the integrals in the elemental equations are evaluated manually. This, however, does not provide a practical procedure for general cases. Therefore, by using Taylor series expansion and truncating after a selected number of terms, we can replace these functions (approximately) with polynomial functions. Substitution of transcendental functions with polynomials allows the use of quadrature in multiscale finite element schemes, making their computer implementation straightforward.

We now consider the expansion of hyperbolic function (Sinh) given as

$$\text{Sinh}\, x = x + \frac{x^3}{3!} + \frac{x^5}{5!} + \frac{x^7}{7!} + \cdots$$

and rewrite it using

$$\begin{cases} x^2 = lx - x(l-x) \\ x^3 = xx^2 = l^2x - lx(l-x) - xx(l-x) \\ x^4 = xx^3 = l^3x - l^2x(l-x) - 2lxx(l-x) + (x(l-x))^2 \\ \vdots \end{cases} \tag{3.39}$$

Higher-order terms can be calculated similarly. Following the outlined technique, bubble-enriched shape functions are found. For example, truncating after the second term, third-order functions are found as

$$N_1 = \frac{(l-x)\left(1 + D_a\frac{(l-x^2)}{6}\right)}{l\left(1+\frac{D_a}{6}h^2\right)} = \frac{l-x}{l} - \frac{x(l-x)(2l-x)}{l\left(\frac{6}{D_a}+l^2\right)} \tag{3.40}$$

and

$$N_2 = \frac{x\left(\left(1+\frac{D_a}{6}x^2\right)\right)}{l\left(1+\frac{D_a}{6}l^2\right)} = \frac{x}{l} - \frac{x(l-x)(l+x)}{l\left(\frac{6}{D_a}+l^2\right)}. \tag{3.41}$$

Figure 3.4 provides a schematic representation of the above described enrichments for linear Lagrangian shape functions.

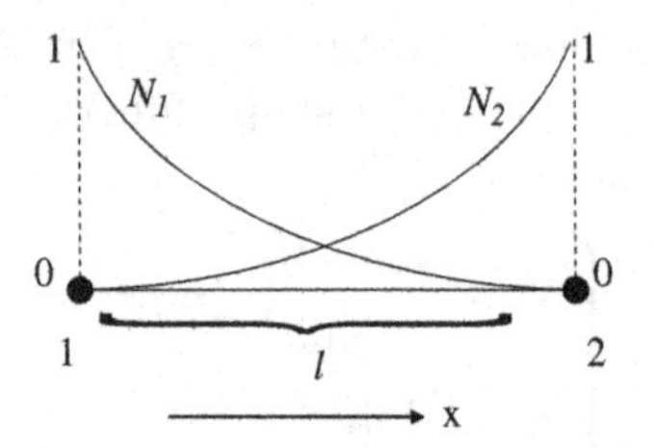

Fig. 3.4 Bubble function-enriched linear Lagrangian shape functions.

Step 3.

Shape functions given in Eqs. (3.40) and (3.41) each consist of a normal Lagrangian shape function part (their first terms) and an enrichment part which is nonzero at the interior of the element and uniformly vanish on its nodes. Here we use ϕ_1 and ϕ_2 to represent these "bubble" functions:

$$\begin{cases} \phi_1 = \dfrac{x(l-x)(2l-x)}{l(\frac{6}{D_a}+l^2)} \\ \phi_2 = \dfrac{x(l-x)(l+x)}{l(\frac{6}{D_a}+l^2)} \end{cases} . \tag{3.42}$$

In the local coordinate system of ξ $(-1,+1)$ these bubble functions are written as

$$\begin{cases} \phi_1 = \dfrac{(1-\xi^2)(3-\xi)}{8\left(1+\frac{6}{D_a l^2}\right)} = b(1-\xi^2)(3-\xi) \\ \phi_2 = \dfrac{(1-\xi^2)(3+\xi)}{8\left(1+\frac{6}{D_a l^2}\right)} = b(1-\xi^2)(3+\xi) \end{cases} , \tag{3.43}$$

where $\xi = 1 - \frac{2x}{l}$ and $b = \frac{1}{8}(1+\frac{6}{D_a l^2})^{-1}$.

Using a similar procedure, fifth-order bubble-enriched linear element can be derived as

$$\begin{cases} \phi_1 = B[b_2(1-\xi^2) + b_3(1-\xi^2)(1-\xi) + b_4(1-\xi^2)^2 \\ \qquad + b_5(1-\xi^2)^2(1-\xi)] \\ \phi_2 = B[b_2(1-\xi^2) + b_3(1-\xi^2)(1+\xi) \\ \qquad + b_4(1-\xi^2)^2 + b_5(1-\xi^2)^2(1+\xi)] \end{cases} , \tag{3.44}$$

in which

$$B = \frac{1}{l\left(1 + \frac{D_a l^2}{3!} + \frac{D_a^2 l^4}{5!}\right)}, \quad b_2 = \left(-\frac{D_a l}{3!} - \frac{D_a^2 l^3}{5!}\right)\frac{l^2}{4},$$

$$b_3 = \left(-\frac{D_a}{3!} - \frac{3D_a^2 l^2}{5!}\right)\frac{l^3}{8}, \quad b_4 = \frac{D_a^2}{5!}\frac{l^5}{8}, \quad b_5 = \left(\frac{D_a^2}{5!} + \frac{6D_a^3 l^2}{7!}\right)\frac{l^5}{32}.$$

We also derive the required coefficients for the seventh- and ninth-order bubble functions as:

Seventh-order

$$B = \frac{1}{l\left(1 + \frac{D_a l^2}{3!} + \frac{D_a^2 l^4}{5!} + \frac{D_a^3 l^6}{7!}\right)}, \quad b_2 = -\left(\frac{D_a l}{3!} + \frac{D_a^2 l^3}{5!} + \frac{D_a^3 l^5}{7!}\right)\frac{l^2}{4},$$

$$b_3 = -\left(\frac{D_a}{3!} + \frac{3D_a^2 l^2}{5!} + \frac{5D_a^3 l^4}{7!}\right)\frac{l^3}{8}, \quad b_4 = \left(\frac{2D_a^2 l}{5!} + \frac{4D_a^3 l^3}{7!}\right)\frac{l^4}{16},$$

$$b_5 = \left(\frac{D_a^2}{5!} + \frac{6D_a^3 l^2}{7!}\right)\frac{l^5}{32}, \quad b_6 = -\left(\frac{3D_a^3 l}{7!}\right)\frac{l^6}{64},$$

$$b_7 = -\left(\frac{D_a^3}{7!}\right)\frac{l^7}{128}.$$

Ninth-order

$$B = \frac{1}{l\left(1 + \frac{D_a l^2}{3!} + \frac{D_a^2 l^4}{5!} + \frac{D_a^3 l^6}{7!} + \frac{D_a^4 l^8}{9!}\right)},$$

$$b_2 = -\left(\frac{D_a l}{3!} + \frac{D_a^2 l^3}{5!} + \frac{D_a^3 l^5}{7!} + \frac{D_a^4 l^7}{9!}\right)\frac{l^2}{4},$$

$$b_3 = -\left(\frac{D_a}{3!} + \frac{3D_a^2 l^2}{5!} + \frac{5D_a^3 l^4}{7!} + \frac{7D_a^4 l^6}{9!}\right)\frac{l^3}{8},$$

$$b_4 = \left(\frac{2D_a^2 l}{5!} + \frac{4D_a^3 l^3}{7!} + \frac{6D_a^4 l^5}{9!}\right)\frac{l^4}{16},$$

$$b_5 = \left(\frac{D_a^2}{5!} + \frac{6D_a^3 l^3}{7!} + \frac{15D_a^4 l^4}{9!}\right)\frac{l^5}{32}, \quad b_6 = -\left(\frac{3D_a^3 l}{7!} + \frac{10D_a^4 l^3}{9!}\right)\frac{l^6}{64},$$

$$b_7 = -\left(\frac{D_a^3}{7!} + \frac{10D_a^4 l^2}{9!}\right)\frac{l^7}{128}, \quad b_8 = \left(\frac{4D_a^4 l}{9!}\right)\frac{l^8}{256}, \quad b_9 = \left(\frac{D_a^4}{9!}\right)\frac{l^9}{512}.$$

Step 4.

Following the normal procedure of the Galerkin finite element method, the elemental stiffness matrix (ESM) for Eq. (3.35) is found as

$$\begin{bmatrix} \int_0^l \left(\frac{d\psi_1}{dx}\frac{dw_1}{dx} + D_a w_1 N_1 \right) dx & \int_0^l \left(\frac{d\psi_2}{dx}\frac{dw_1}{dx} + D_a w_1 N_2 \right) dx \\ \int_0^l \left(\frac{d\psi_1}{dx}\frac{dw_2}{dx} + D_a w_2 N_1 \right) dx & \int_0^l \left(\frac{d\psi_2}{dx}\frac{dw_2}{dx} + D_a w_2 N_2 \right) dx \end{bmatrix}. \tag{3.45}$$

An important point about this derivation is to note that the use of bubble-enriched elements does not affect the second-order term (i.e. the Laplacian operator). To show this, we consider the variational statement of the described problem written as

$$(\nabla T, \nabla v) + (D_a T, v) = (f, v). \tag{3.46}$$

Substitution from Eq. (3.12) gives

$$(\nabla T_c, \nabla v) + (\nabla T_b, \nabla v) + (D_a T, v) = (f, v). \tag{3.47}$$

For a linear test function v (i.e. weight function) according to Green's theorem (Franca and Farhat, 1995) we have

$$(\nabla v, \nabla \phi)_{\Omega_e} = -(\Delta v, \phi)_{\Omega_e} + (\nabla v, \phi)_{\Gamma_e} = 0, \tag{3.48}$$

where ϕ is the bubble function. Therefore, Eq. (3.47) is reduced to

$$(\nabla T_c, \nabla v) + (D_a T, v) = (f, v). \tag{3.49}$$

As demonstrated, the bubble function does not affect the Laplacian term of the DR equation.

Example 1. Solution of the DR equation using RFB method.

Consider the following dimensionless DR equation and its associated boundary conditions in one dimension as

$$\begin{cases} \frac{d^2T^*}{dx^{*2}} - D_a T^* = f^* \\ T^* = 0 \quad \text{at } x^* = 0 \\ T^* = 1 \quad \text{at } x^* = 1 \end{cases}. \tag{3.50}$$

In this example $f^* = 0$. After the discretization of the solution domain into 10 elements of equal size (i.e. $l = 0.1$) for a value of $D_a = 500$, the standard

Galerkin finite element scheme gives the following ESM

$$\text{ESM} = \begin{bmatrix} 26.6667 & -1.6667 \\ -1.6667 & 26.6667 \end{bmatrix},$$

and after the assembly of ESMs over the common nodes the global stiffness matrix is found as

$$\begin{bmatrix} 26.67 & -1.67 & & & & & & & & & \\ -1.67 & 53.33 & -1.67 & & & & & & & \mathbf{0} & \\ 0 & -1.67 & 53.33 & -1.67 & & & & & & & \\ & & -1.67 & 53.33 & -1.67 & & & & & & \\ & & & -1.67 & 53.33 & -1.67 & & & & & \\ & & & & -1.67 & 53.33 & -1.67 & & & & \\ & & & & & -1.67 & 53.33 & -1.67 & & & \\ & & & & & & -1.67 & 53.33 & -1.67 & & \\ & & & & & & & -1.67 & 53.33 & -1.67 & 0 \\ & \mathbf{0} & & & & & & & -1.67 & 53.33 & -1.67 \\ & & & & & & & & & -1.67 & 26.67 \end{bmatrix}$$

In the exact RFB method, the shape and weight functions for the linear Lagrangian element are

$$\psi_1 = w_1 = 1 - \frac{x}{l}, \quad N_1 = \frac{\sinh\sqrt{D_a}(l-x)}{\sinh\sqrt{D_a}l},$$

$$\psi_2 = w_2 = \frac{x}{l}, \quad N_2 = \frac{\sinh\sqrt{D_a}x}{\sinh\sqrt{D_a}l}$$

Therefore, the ESM is found as

$$\text{ESM} = \begin{bmatrix} \sqrt{D_a}\dfrac{\cosh\sqrt{D_a}l}{\sinh\sqrt{D_a}l} & \dfrac{\sqrt{D_a}}{\sinh\sqrt{D_a}l} \\ \dfrac{\sqrt{D_a}}{\sinh\sqrt{D_a}l} & \sqrt{D_a}\dfrac{\cosh\sqrt{D_a}l}{\sinh\sqrt{D_a}l} \end{bmatrix}. \tag{3.51}$$

Insertion of $D_a = 500$ (note that in this case the DR equation is physically representing a dissipative reaction corresponding to a positive D_a), and $l = 0.1$ gives

$$\text{ESM} = \begin{bmatrix} 22.8774 & -4.8350 \\ -4.8350 & 22.8774 \end{bmatrix},$$

and the global matrix corresponding to the ESM found in this case is

$$\begin{bmatrix}
45.7549 & -4.8350 & & & & & & & \\
-4.8350 & 45.7549 & -4.8350 & & & & & \text{\Huge 0} & \\
0 & -4.8350 & 45.7549 & -4.8350 & & & & & \\
 & & -4.8350 & 45.7549 & -4.8350 & & & & \\
 & & & -4.8350 & 45.7549 & -4.8350 & & & \\
 & & & & -4.8350 & 45.7549 & -4.8350 & & \\
 & & & & & -4.8350 & 45.7549 & -4.8350 & 0 \\
\text{\Huge 0} & & & & & & -4.8350 & 45.7549 & -4.8350 \\
 & & & & & & & -4.8350 & 45.7549
\end{bmatrix}.$$

After the calculation of global stiffness matrices in each of the above described cases, the resulting set of algebraic equations are solved. The results of these solutions are compared in Fig. 3.5. The standard Galerkin finite element solution based on ordinary Lagrangian elements is stable but comparison of its results with the analytical (exact) solution shows that it is inaccurate. In contrast, the exact RFB solution is stable and accurate, and produces answers that are identical to the exact solution. To further clarify this point all three sets of numerical results are compared in Table 3.1.

It should be noted that the accuracy of the results generated by the standard Galerkin finite element scheme can be improved using successive mesh refinements. In realistic cases, however, the required level of mesh refinement may be very high, precluding such a course of action.

Example 2. DR equation — exact RFB, $D_a = 500$ and $l = 0.2$.

In this example the element length is increased to $l = 0.2$ and, therefore, the domain is discretized into a coarser mesh of five elements of equal size. In this case the ESM corresponding to the exact RFB method with $D_a = 500$

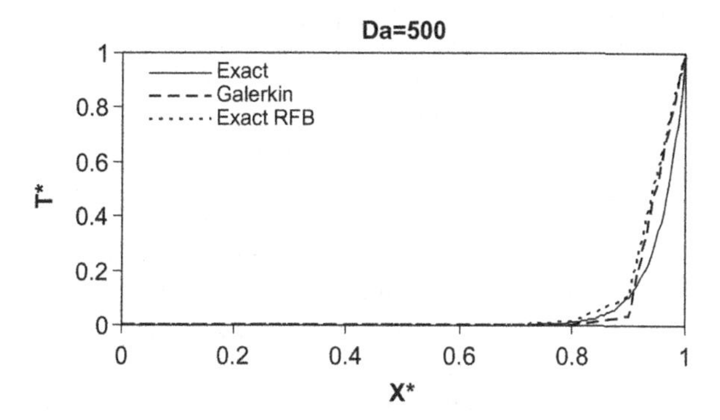

Fig. 3.5 Results obtained by the standard Galerkin, analytical, and exact RFB function methods ($D_a = 500$ and $l = 0.1$).

Table 3.1. Comparison of numerical results at nodal points ($D_a = 500$ and $l = 0.1$).

X	Exact	Exact RFB	Standard Galerkin
0.0	0	0	0
0.1	0	0	0
0.2	0	0	0
0.3	0	0	0
0.4	0	0	0
0.5	0	0	0
0.6	0.0001	0.0001	0
0.7	0.0012	0.0012	0
0.8	0.0114	0.0114	0.001
0.9	0.1069	0.1069	0.0313
1.0	1	1	1

and $l = 0.2$ is found as

$$\mathrm{ESM} = \begin{bmatrix} 22.3665 & -0.5109 \\ -0.5109 & 22.3665 \end{bmatrix},$$

and the global stiffness matrix is thereby calculated as

$$\begin{bmatrix} 44.733 & -0.5109 & 0 & 0 \\ -0.5109 & 44.733 & -0.5109 & 0 \\ 0 & -0.5109 & 44.733 & -0.5109 \\ 0 & 0 & -0.5109 & 44.733 \end{bmatrix}.$$

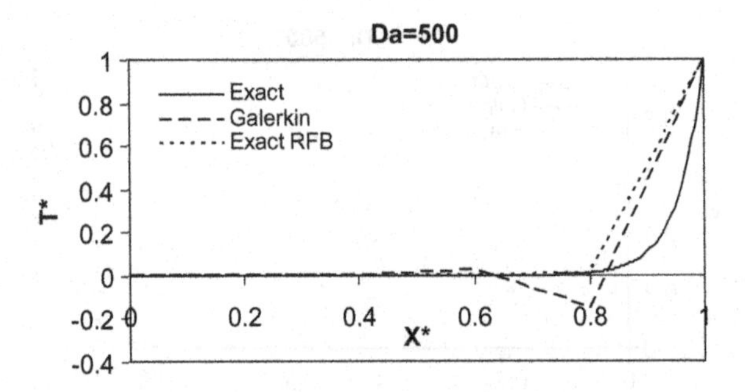

Fig. 3.6 Comparison of the results obtained by the standard Galerkin, analytical, and exact RFB function methods ($D_a = 500$ and $l = 0.2$).

Table 3.2. Comparison of numerical results at nodal points ($D_a = 500$ and $l = 0.2$).

X	Exact	Exact RFB	Standard Galerkin
0.0	0	0	0
0.2	0	0	0.0367
0.4	0	0	−0.0959
0.6	0.0001	0.0001	0.2142
0.8	0.0114	0.0114	−0.4642
1.0	1	1	1

Figure 3.6 clearly shows that such a simple change, combined with the use of a coarser mesh, results in the failure of the standard Galerkin method to preserve the stability of the solution. In contrast the exact RFB method again generates results that are identical with the exact solution (Table 3.2).

As shown in Examples 1 and 2, treatment of multiscale behavior via standard finite element schemes becomes more problematic using a coarser mesh. This adversely affects the cost of computations based on such methods in realistic applications. In addition to the influence of the degree of mesh refinement, we can check the effects of physical parameters such as D_a on the multiscale behavior of the prescribed problem.

Example 3. DR equation — exact RFB, $D_a = 1000$ and $l = 0.1$.

Using solution procedures identical to Examples 1 and 2, results obtained for this example for the standard Galerkin finite element, exact RFB, and analytical solution are shown in Fig. 3.7. Additionally, numerical results generated by the analytical and exact RFB methods are shown in Table 3.3.

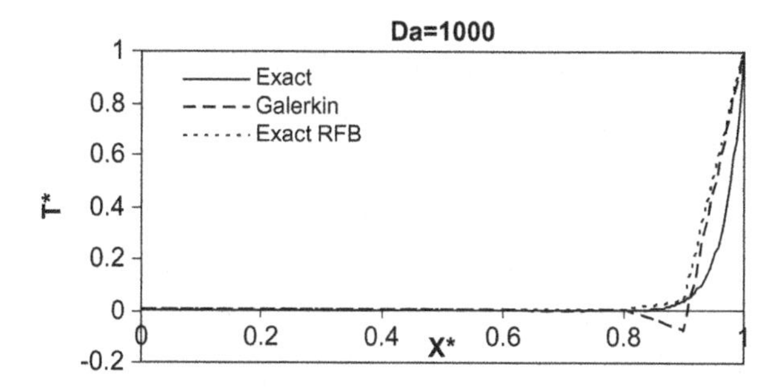

Fig. 3.7 Comparison of the results obtained by the standard Galerkin, analytical, and exact RFB function methods ($D_a = 1000$ and $l = 0.1$).

Table 3.3. Comparison of numerical values of RFB method at nodal points ($D_a = 1000$ and $l = 0.1$).

X	Exact	Exact RFB
0.0	0	0
0.1	0	0
0.2	0	0
0.3	0	0
0.4	0	0
0.5	0	0
0.6	0	0
0.7	0.0001	0.0001
0.8	0.0018	0.0018
0.9	0.0423	0.0423
1.0	1	1

As shown in Fig. 3.7 the standard Galerkin method based on ordinary Lagrangian elements has again generated unstable results. On the other hand, the exact RFB scheme has generated a super-convergent solution.

Example 4. DR equation — exact RFB, $D_a = 500$, $l = 0.1$, and $f^* = 100$.

We now consider the solution of a nonhomogeneous DR equation with $D_a = 500$, $l = 0.1$ and $f^* = 100$. The ESM obtained using the exact RFB method is

$$\text{ESM} = \begin{bmatrix} 22.8774 & -4.835 \\ -4.835 & 22.8774 \end{bmatrix},$$

which is identical to the one calculated in Example 1. Therefore the global stiffness equation also remains similar to the one found in Example 1. However, as the equation is nonhomogeneous the load vector is different and should be recalculated.

Calculation of the load vector:

To find the load vector Eq. (3.29) is solved and resulting ϕ_f is inserted into Eq. (3.30). Equation (3.30) is then used to construct the weak form of Eq. (3.35).
Therefore,

$$T = \sum_{i=1}^{n} T_i(\psi_i + \phi_i) + \phi_f.$$

Using the above definition of T in conjunction with linear Lagrangian shape functions we have

$$\int_0^l \left(\frac{d\psi_1}{dx}\frac{dw_1}{dx}T_1 + \frac{d\psi_2}{dx}\frac{dw_1}{dx}T_2 + \frac{d\phi_f}{dx}\frac{dw_1}{dx} + D_a(w_1 N_1 T_1 + w_1 N_2 T_2 + w_1\phi_f) + f^* w_1 \right) dx = -w_1\Phi\Big|_{\Gamma_e},$$

$$\int_0^l \left(\frac{d\psi_1}{dx}\frac{dw_2}{dx}T_1 + \frac{d\psi_2}{dx}\frac{dw_2}{dx}T_2 + \frac{d\phi_f}{dx}\frac{dw_2}{dx} + D_a(w_2 N_1 T_1 + w_2 N_2 T_2 + w_2\phi_f) + f^* w_2 \right) dx = -w_2\Phi\Big|_{\Gamma_e}.$$

The elemental form can therefore be written as

$$\begin{bmatrix} \int_0^l \left(\frac{d\psi_1}{dx}\frac{dw_1}{dx} + D_a w_1 N_1 \right) dx & \int_0^l \left(\frac{d\psi_2}{dx}\frac{dw_1}{dx} + D_a w_1 N_2 \right) dx \\ \int_0^l \left(\frac{d\psi_1}{dx}\frac{dw_2}{dx} + D_a w_2 N_1 \right) dx & \int_0^l \left(\frac{d\psi_2}{dx}\frac{dw_2}{dx} + D_a w_2 N_2 \right) dx \end{bmatrix} \begin{bmatrix} T_1 \\ T_2 \end{bmatrix}$$

$$= - \begin{bmatrix} w_1\Phi|_{\Gamma_e} + \int_0^l \left(\frac{d\phi_f}{dx}\frac{dw_1}{dx} + D_a w_1 \phi_f + f^* w_1 \right) dx \\ w_2\Phi|_{\Gamma_e} + \int_0^l \left(\frac{d\phi_f}{dx}\frac{dw_2}{dx} + D_a w_2 \phi_f + f^* w_2 \right) dx \end{bmatrix}. \tag{3.52}$$

The ESM (left-hand side of the above equation) remains the same as the homogeneous equation (see Eq. 3.45). Boundary line terms are eliminated because of inter-element connectivity during the assembly and do not appear in the global formulation (these terms are of equal magnitude and opposite signs over common nodes of connected elements). After the completion of the described procedure, the global load vector corresponding to this example is found as

$$B = \begin{bmatrix} -7.217 \\ -7.217 \\ -7.217 \\ -7.217 \\ -7.217 \\ -7.217 \\ -7.217 \\ -7.217 \\ -2.382 \end{bmatrix}.$$

The solution of the nonhomogeneous problem posed in Example 4 via the standard Galerkin finite element, exact RFB, and analytical method are shown in Fig. 3.8. As shown in Table 3.4, exact RFB method generates identical results to the analytical solution in this case.

The exact RFB method used in the previous examples generates identical results to the analytical solution at the nodal points of a computational mesh. However, as the development of an exact method depends on manual integration of terms of the elemental stiffness equation (such as the terms seen in Eq. 3.51), it may not be possible to construct such a scheme for practical problems under realistic conditions. As mentioned earlier, the semi-discrete method (SD method), developed by Parvazinia *et al.* (2006b), which uses Taylor series expansion of the analytical shape functions, can

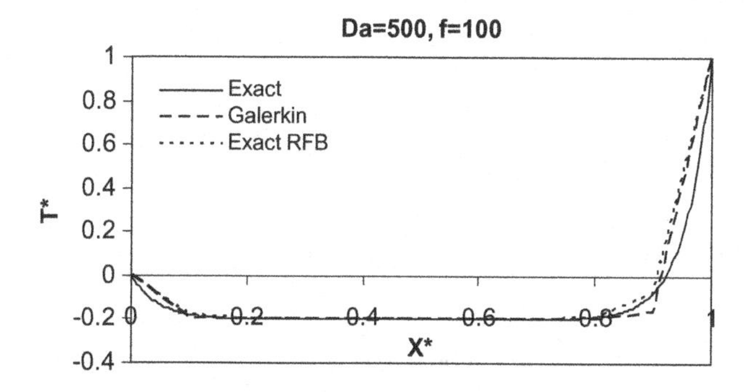

Fig. 3.8 Comparison of the results obtained by the standard Galerkin, analytical, and exact RFB function methods ($D_a = 500$, $l = 0.1$, and $f^* = 100$).

Table 3.4. Comparison of numerical results of RFB and analytical methods at nodal points (D_a = 500, l = 0.1, and f^* = 100).

X	Exact	Exact RFB
0.0	0	0
0.1	−0.1786	−0.1786
0.2	−0.1977	−0.1977
0.3	−0.1998	−0.1998
0.4	−0.2000	−0.2000
0.5	−0.2000	−0.2000
0.6	−0.1998	−0.1998
0.7	−0.1985	−0.1985
0.8	−0.1863	−0.1863
0.9	−0.0717	−0.0717
1.0	1	1

be used to develop effective RFB schemes for practical problems (despite the loss of exactness in the bubble-enriched shape functions). To demonstrate the possibilities offered by this approach, we now consider a number of examples.

Example 5. DR equation — $D_a = 500$ and $l = 0.1$, third-order RFB.

Consider the following shape and weight functions associated with a linear Lagrangian element enriched by the third-order polynomial found via Taylor series expansion of the original functions obtained by the analytical solution of the governing DR equation.

$$\psi_1 = w_1 = 1 - \frac{x}{l}, \quad N_1 = 1 - \frac{x}{l} - \frac{x(l-x)(2l-x)}{l\left(\frac{6}{D_a} + l^2\right)}, \tag{3.53a}$$

$$\psi_2 = w_2 = \frac{x}{l}, \quad N_2 = \frac{x}{l} - \frac{x(l-x)(l+x)}{l\left(\frac{6}{D_a} + l^2\right)}. \tag{3.53b}$$

Using these functions, the ESM is found as

$$\text{ESM} = \begin{bmatrix} \frac{1}{l} + \frac{D_a}{3}l + 1.0667bD_a & -\frac{1}{l} + \frac{D_a}{6}l + 1.0667bD_a \\ -\frac{1}{l} + \frac{D_a}{6}l + 1.0667bD_a & \frac{1}{l} + \frac{D_a}{3}l + 1.0667bD_a \end{bmatrix}$$

in which

$$b = \frac{1}{8}\left(1 + \frac{6}{D_a l^2}\right)^{-1}. \tag{3.54}$$

Comparison of the members of the above matrix with their simpler counterparts, which appear using the standard Galerkin scheme based on ordinary linear Lagrangian elements, shows that these terms are "enriched" by the addition of $1.0667bD_a$ to the simpler original terms. Furthermore, this level of "compensation" remains unchanged for all of the elements in the computational mesh. After the insertion of $D_a = 500, l = 0.1$ into Eq. (3.54), b is found to be 0.0568 and hence the ESM in this case can be written as

$$\text{ESM} = \begin{bmatrix} 23.6372 & -4.6961 \\ -4.6961 & 23.6372 \end{bmatrix}.$$

The global matrix found after the assembly of the elemental matrices is shown below. Similar to the standard Galerkin method the global matrix obtained using third-order RFB functions is symmetric

$$\begin{bmatrix}
47.2745 & -4.6961 & & & & & & & & \\
-4.6961 & 47.2745 & -4.6961 & & & & & \mathbf{0} & & \\
0 & -4.6961 & 47.2745 & -4.6961 & & & & & & \\
 & & -4.6961 & 47.2745 & -4.6961 & & & & & \\
 & & & -4.6961 & 47.2745 & -4.6961 & & & & \\
 & & & & -4.6961 & 47.2745 & -4.6961 & & & \\
 & & & & & -4.6961 & 47.2745 & -4.6961 & 0 & \\
 & \mathbf{0} & & & & & -4.6961 & 47.2745 & -4.6961 & \\
 & & & & & & & -4.6961 & 47.2745 &
\end{bmatrix}.$$

After the calculation of the global stiffness matrix, the system of algebraic equations arising from the application of third-order RFB method to the DR equation can be carried out. Results of this solution are shown in Fig. 3.9 and compared with corresponding results obtained using analytical and the standard Galerkin finite element methods. As Fig. 3.9 shows, the third-order RFB is capable of generating stable-accurate solution. This solution is slightly better than the result generated by the standard Galerkin

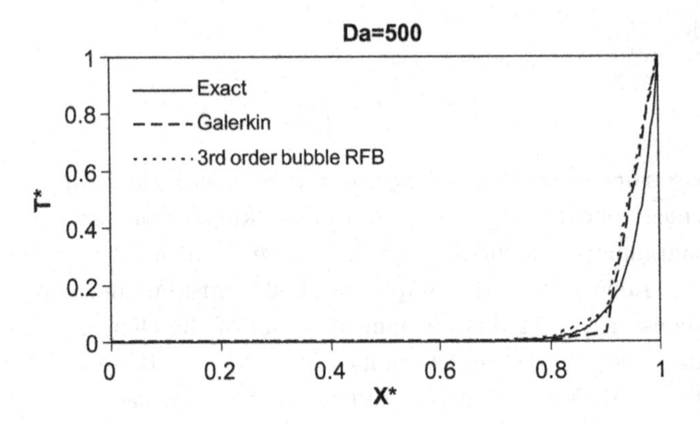

Fig. 3.9 Comparison of the results obtained by the standard Galerkin, analytical, and third-order RFB function methods ($D_a = 500$ and $l = 0.1$).

Table 3.5. Comparison of results generated by the third-order bubble functions with analytical solution ($D_a = 500$ and $l = 0.1$).

X	Exact	Third-order bubble
0.0	0	0
0.1	0	0
0.2	0	0
0.3	0	0
0.4	0	0
0.5	0	0
0.6	0.0001	0.0001
0.7	0.0012	0.0010
0.8	0.0114	0.0101
0.9	0.1069	0.1003
1.0	1	1

scheme using ordinary elements. However, as shown in Table 3.5, in this case the numerical results obtained using the third-order RFB are not totally accurate.

Example 6. DR equation — $D_a = 500$ and $l = 0.2$.

For further evaluation of the performance of the described third-order bubble functions and, in particular, to compare its accuracy with the standard Galerkin method based on ordinary elements, we repeat the solution

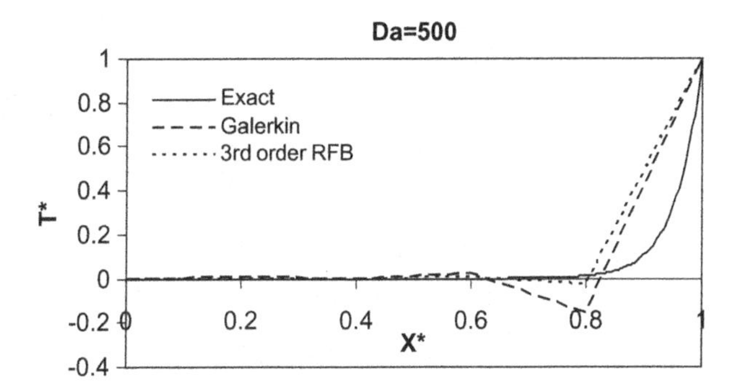

Fig. 3.10 Comparison of the results obtained by the standard Galerkin, analytical, and third-order RFB function methods ($D_a = 500$ and $l = 0.2$).

of the DR equation, this time using longer elements of length $l = 0.2$ (i.e. using a mesh twice coarser than previously). From Eq. (3.54) the bubble coefficient in this case is found to be $b = 0.0962$.

In Fig. 3.10 the results obtained for this case are shown and compared with the analytical solution and the results produced by the standard Galerkin finite element method.

Figure 3.10 clearly shows the difference between the results generated by the standard Galerkin scheme using ordinary elements and the bubble-enriched solution. Although third-order RFB has failed to generate accurate results, they are stable and closer to the analytical solution than the solution generated by the standard Galerkin method.

Example 7. DR equation — exact RFB, $D_a = -500$ and $l = 0.1$.

We now consider a productive reaction case represented by the DR equation by choosing a negative value for the Damköhler number as $D_a = -500$. Here a mesh of 10 elements of equal length is used, i.e. $l = 0.1$. In this case we again compare the solutions obtained via the analytical, standard Galerkin based on ordinary elements and bubble-enriched finite element methods. The elemental and global stiffness matrices calculated in this case are:

(i) *Galerkin method based on ordinary elements*

$$\mathrm{ESM} = \begin{bmatrix} -6.6667 & -18.3333 \\ -18.3333 & -6.6667 \end{bmatrix}$$

and

$$\begin{bmatrix}
-13.3333 & -18.3333 & & & & & & & \\
-18.3333 & -13.3333 & -18.3333 & & & & & & \mathbf{0} \\
0 & -18.3333 & -13.3333 & -18.3333 & & & & & \\
 & & -18.3333 & -13.3333 & -18.3333 & & & & \\
 & & & -18.3333 & -13.3333 & -18.3333 & & & \\
 & & & & -18.3333 & -13.3333 & -18.3333 & & \\
 & & & & & -18.3333 & -13.3333 & -18.3333 & 0 \\
\mathbf{0} & & & & & & -18.3333 & -13.3333 & -18.3333 \\
 & & & & & & & -18.3333 & -13.3333
\end{bmatrix}$$

(ii) *Exact RFB*:

$$\text{ESM} = \begin{bmatrix} -17.5439 & -28.4216 \\ -28.4216 & -17.5439 \end{bmatrix}$$

and

$$\begin{bmatrix}
-35.0878 & -28.4216 & & & & & & & \\
-28.4216 & -35.0878 & -28.4216 & & & & & & \mathbf{0} \\
0 & -28.4216 & -35.0878 & -28.4216 & & & & & \\
 & & -28.4216 & -35.0878 & -28.4216 & & & & \\
 & & & -28.4216 & -35.0878 & -28.4216 & & & \\
 & & & & -28.4216 & -35.0878 & -28.4216 & & \\
 & & & & & -28.4216 & -35.0878 & -28.4216 & 0 \\
\mathbf{0} & & & & & & -28.4216 & -35.0878 & -28.4216 \\
 & & & & & & & -28.4216 & -35.0878
\end{bmatrix}$$

Solutions generated by the above methods are shown in Fig. 3.11. The exact FRB scheme has clearly generated a more accurate solution than the Galerkin scheme using ordinary elements.

In contrast with the case of $D_a = 500$ (Fig. 3.5), where the analytical solution is an exponential function, the standard Galerkin method has

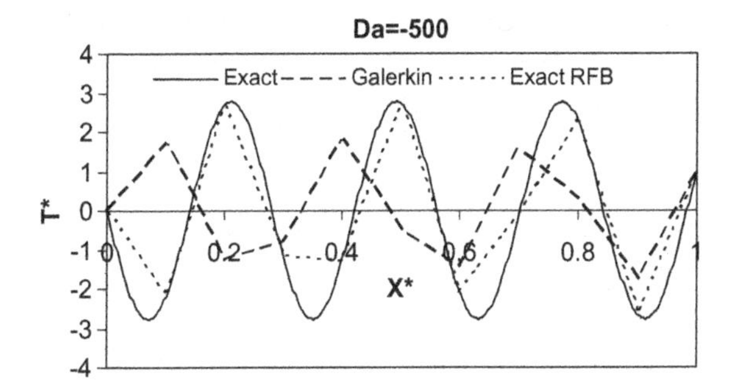

Fig. 3.11 Comparison of the results obtained by the standard Galerkin, analytical, and exact RFB function methods ($D_a = -500$ and $l = 0.1$).

Table 3.6. Comparison of numerical results generated by the exact RFB method with the analytical solution at nodal points ($Da = -500$ and $l = 0.1$).

X	Exact	Exact RFB
0.0	0	0
0.1	−2.1783	−2.1783
0.2	2.6892	2.6892
0.3	−1.1416	−1.1416
0.4	−1.2798	−1.2798
0.5	2.7216	2.7216
0.6	−2.0801	−2.0801
0.7	−0.1536	−0.1536
0.8	2.2697	2.2697
0.9	−2.6485	−2.6485
1.0		1

generated a significantly less accurate result. This is due to the oscillatory nature of the solution in this case and inability of the standard Galerkin technique to generate accurate results for such cases without significant mesh refinements. As shown in Table 3.6, the nodal results obtained by the exact RFB and analytical solutions in this case are identical.

Example 8. DR equation — exact RFB, $D_a = -500$ and $l = 0.2$.

In comparison with Example 2 (which represented a dissipative reaction case), in this example a productive reaction case corresponding to

$D_a = -500$ is simulated using a coarse mesh of five equal length elements, i.e. $l = 0.2$. The ESM obtained the exact RFB method in this case is

$$\text{ESM} = \begin{bmatrix} 5.4780 & 23.0219 \\ 23.0219 & 5.4780 \end{bmatrix},$$

and the corresponding global matrix is calculated as

$$\begin{bmatrix} 10.9561 & 23.0219 & 0 & 0 \\ 23.0219 & 10.9561 & 23.0219 & 0 \\ 0 & 23.0219 & 10.9561 & 23.0219 \\ 0 & 0 & 23.0219 & 10.9561 \end{bmatrix}.$$

Using the above matrix, the solution of the problem described in this example is obtained. Comparison of the results generated by the standard Galerkin, exact FRB and analytical methods in Fig. 3.12, shows the total failure of the standard Galerkin method when the present coarser mesh is employed. In contrast, the nodal values found by the exact FRB method remain identical to the analytical solution (Table 3.7).

Example 9. DR equation — exact RFB, $D_a = -10000$ and $l = 0.2$.

We now consider a significantly more difficult situation represented by the DR equation for a productive reaction case corresponding to a $D_a = -10000$. A coarse mesh is again used in which $l = 0.2$. In this case the ESM corresponding to the exact RFB method is found as

$$\text{ESM} = \begin{bmatrix} 44.6995 & -109.5356 \\ -109.5356 & 44.6995 \end{bmatrix},$$

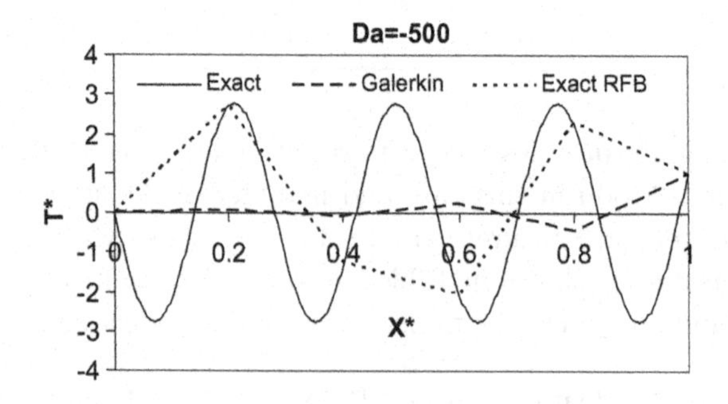

Fig. 3.12 Comparison of the results obtained by the standard Galerkin, analytical, and exact RFB function methods ($D_a = -500$ and $l = 0.2$).

Table 3.7. Comparison of numerical results generated by the exact RFB method with analytical solution at nodal points ($D_a = -500$ and $l = 0.2$).

X	Exact	Exact RFB
0.0	0	0
0.2	2.6892	2.6892
0.4	−1.2798	−1.2798
0.6	−2.0801	−2.0801
0.8	2.2697	2.2697
1.0	1	1

and hence the global matrix is calculated as

$$\begin{bmatrix} 89.3990 & -109.5356 & 0 & 0 \\ -109.5356 & 89.3990 & -109.5356 & 0 \\ 0 & -109.5356 & 89.3990 & -109.5356 \\ 0 & 0 & -109.5356 & 89.3990 \end{bmatrix}.$$

Using the data represented in the above matrix, the solution based on bubble function-enriched elements in this case is found and compared with solutions generated by the standard Galerkin and analytical methods. As expected, the standard Galerkin scheme based on ordinary elements fails to generate a stable solution whilst the nodal values obtained by the exact RFB method remain identical to the analytical solution (Table 3.8).

Example 10. DR equation — exact RFB, $D_a = -500$, $l = 0.1$, and $f^* = 100$.

Table 3.8. Comparison of numerical results generated by the exact RFB method with analytical solution at nodal points ($D_a = -10000$ and $l = 0.2$).

X	Exact	Exact RFB
0.0	0	0
0.2	−1.8029	−1.8029
0.4	−1.4715	−1.4715
0.6	0.602	0.602
0.8	1.9628	1.9628
1.0	1	1

We now consider a nonhomogeneous DR equation representing a productive reaction case in which $D_a = -500$ and $f^* = 100$. A mesh consisting of 10 elements of equal size is used and hence $l = 0.1$. Similar to the homogeneous case, the ESM is calculated using Eq. (3.51) and is found as

$$\text{ESM} = \begin{bmatrix} -17.5439 & -28.4216 \\ -28.4216 & -17.5439 \end{bmatrix}.$$

After the assembly of the elemental equations the global stiffness matrix is found as

$$\begin{bmatrix} -35.0878 & -28.4216 & & & & & & & \\ -28.4216 & -35.0878 & -28.4216 & & & & & \mathbf{0} & \\ 0 & -28.4216 & -35.0878 & -28.4216 & & & & & \\ & -28.4216 & -35.0878 & -28.4216 & & & & & \\ & & -28.4216 & -35.0878 & -28.4216 & & & & \\ & & & -28.4216 & -35.0878 & -28.4216 & & & \\ & & & & -28.4216 & -35.0878 & -28.4216 & 0 & \\ \mathbf{0} & & & & & -28.4216 & -35.0878 & -28.4216 & \\ & & & & & & -28.4216 & -35.0878 & \end{bmatrix},$$

and the corresponding load vector is

$$\begin{bmatrix} -18.3862 \\ -18.3862 \\ -18.3862 \\ -18.3862 \\ -18.3862 \\ -18.3862 \\ -18.3862 \\ -18.3862 \\ 10.0354 \end{bmatrix}.$$

Using the above data, the required solution is carried out. Results obtained by the solution of the DR equation in this case is shown in Fig. 3.13 and compared with the analytical results and standard Galerkin solution generated using ordinary elements. As shown in Table 3.9, the exact RFB method again generates perfect solutions at nodal points.

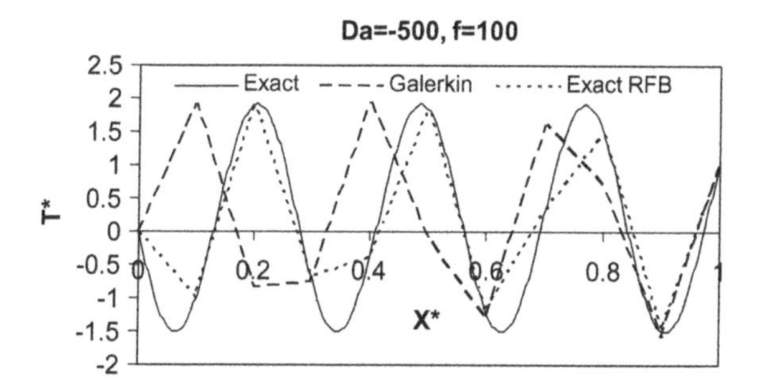

Fig. 3.13 Comparison of numerical results generated by the exact RFB method with analytical solution at nodal points ($D_a = -500$, $l = 0.1$, and $f^* = 100$).

Table 3.9. Comparison of numerical results of exact RFB and analytical methods at nodal points ($D_a = -500$, $l = 0.1$, and $f^* = 100$).

X	Exact	Exact RFB
0.0	0	0
0.1	−1.0129	−1.0129
0.2	1.8974	1.8974
0.3	−0.6826	−0.6826
0.4	−0.4078	−0.4078
0.5	1.833	1.833
0.6	−1.2082	−1.2082
0.7	0.3055	0.3055
0.8	1.4779	1.4779
0.9	−1.4832	−1.4832
1.0	1	1

3.3.2 *Multiscale finite element solution of the CD equation using RFB function method*

Steady state CD equation can be written in a dimensionless form as

$$\left(P_e \frac{\partial T^*}{\partial x^*}\right) - \left(\frac{\partial^2 T^*}{\partial x^{*2}}\right) = f^*, \tag{3.55}$$

where P_e is the Peclet number. Node-based solution of Eq. (3.55) provides the "analytical" shape functions N_1 and N_2 and after the substitution of

T^* by N_1 and N_2, we have

$$\begin{cases} -\dfrac{d^2N_1}{dx^2} + P_e\dfrac{dN_1}{dx} = 0 \quad \text{for } x \in [0-l] \\ N_1 = \psi_1 \Rightarrow \begin{cases} N_1(0) = 1 \\ N_1(l) = 0 \end{cases} \end{cases} \tag{3.56}$$

and

$$\begin{cases} -\dfrac{d^2N_2}{dx^2} + P_e\dfrac{dN_2}{dx} = 0 \quad \text{for } x \in [0-l] \\ N_2 = \psi_2 \Rightarrow \begin{cases} N_2(0) = 0 \\ N_2(l) = 1 \end{cases} \end{cases}, \tag{3.57}$$

where l is a characteristic element length and ψ_i is a linear shape function. Solution of the above equations yields the bubble shape functions which, using a local elemental coordinate system, can be expressed as

$$\begin{cases} N_1 = \dfrac{\exp(p_e x) - \exp(p_e l)}{1 - \exp(p_e l)} \\ N_2 = \dfrac{1 - \exp(p_e x)}{1 - \exp(p_e l)} \end{cases}. \tag{3.58}$$

Taylor series expansion of the exponential function e^x gives

$$e^x = 1 + x + \frac{x^2}{2!} + \frac{x^3}{3!} + \cdots. \tag{3.59}$$

Similar to the solution of the DR equation, the following forms can now be used

$$\begin{cases} x = l - (l - x) \\ x^2 = lx - x(l - x) \\ x^3 = xx^2 = l^2x - lx(l - x) - xx(l - x) \\ x^4 = xx^3 = l^3x - l^2x(l - x) - 2lxx(l - x) + (x(l - x))^2 \\ \vdots \end{cases} \tag{3.60}$$

After the selection of a local coordinate system as $\xi(-1, +1)$ where $x = \frac{l}{2}(1 + \xi)$, the second-order truncated bubble function shape functions are found as

$$\begin{cases} N_1 = \dfrac{1}{2}(1-\xi) + b(1-\xi^2) \\ N_2 = \dfrac{1}{2}(1+\xi) - b(1-\xi^2) \end{cases}, \tag{3.61}$$

where

$$b = \frac{lp_e}{8(1+0.5p_e l)}, \tag{3.62}$$

where l is a characteristic element length. The sign of bubble coefficient is different for the beginning and end nodes in the two-node linear element selected here. This is because the slopes of the bubble functions at the inter-element boundaries remain unchanged. The opposite signs of the bubble functions affect the convection term in the CD equation.

If the first five terms in the Taylor series expansion are kept we obtain the following fourth-order bubble shape functions

$$\begin{cases} N_1 = \dfrac{1}{2}(1-\xi) + b_1(1-\xi^2) + b_2(1+\xi)(1-\xi^2) + b_3(1-\xi^2)^2 \\ N_2 = \dfrac{1}{2}(1+\xi) + b_1(1-\xi^2) + b_2(1+\xi)(1-\xi^2) + b_3(1-\xi^2)^2 \end{cases}, \tag{3.63}$$

where

$$b_1 = \frac{\frac{1}{2!}p_e^2 l^2 + \frac{1}{3!}p_e^3 l^3 + \frac{1}{4!}p_e^4 l^4}{4\left(p_e l + \frac{p_e^2 l^2}{2!} + \frac{p_e^3 l^3}{3!} + \frac{p_e^4 l^4}{4!}\right)},$$

$$b_2 = \frac{\frac{1}{3!}p_e^3 l^3 + \frac{2}{4!}p_e^4 l^4}{8\left(p_e l + \frac{p_e^2 l^2}{2!} + \frac{p_e^3 l^3}{3!} + \frac{p_e^4 l^4}{4!}\right)},$$

$$b_3 = \frac{\frac{1}{4!}p_e^4 l^4}{16\left(p_e l + \frac{p_e^2 l^2}{2!} + \frac{p_e^3 l^3}{3!} + \frac{p_e^4 l^4}{4!}\right)},$$

where l is a characteristic element length.

The general form of the elemental stiffness equation corresponding to the CD equation can be derived as

$$\text{ESM} = \begin{bmatrix} \displaystyle\int_0^\ell \left(\frac{d\psi_1}{dx}\frac{dw_1}{dx} + P_e w_1 \frac{dN_1}{dx}\right) dx & \displaystyle\int_0^\ell \left(\frac{d\psi_2}{dx}\frac{dw_1}{dx} + P_e w_1 \frac{dN_2}{dx}\right) dx \\ \displaystyle\int_0^\ell \left(\frac{d\psi_1}{dx}\frac{dw_2}{dx} + P_e w_2 \frac{dN_1}{dx}\right) dx & \displaystyle\int_0^\ell \left(\frac{d\psi_2}{dx}\frac{dw_2}{dx} + P_e w_2 \frac{dN_2}{dx}\right) dx \end{bmatrix}. \tag{3.64}$$

we now consider the solution of typical examples.

Example 11. CD equation — standard Galerkin solution, $P_e = 10$ and $l = 0.1$.

The governing equation and its corresponding boundary conditions used in this example are given as

$$\begin{cases} \left(P_e \dfrac{\partial T^*}{\partial x^*}\right) - \left(\dfrac{\partial^2 T^*}{\partial x^{*2}}\right) = f^* \\ T^* = 0 \quad \text{at } x^* = 0 \\ T^* = 1 \quad \text{at } x^* = 1 \end{cases}.$$

After the discretization of the problem domain into 10 equal size elements for the homogeneous CD equation (i.e. $f^* = 0$) the numerical values of the members of the ESM corresponding to the standard Galerkin method based on ordinary linear elements are found as

$$\text{ESM} = \begin{bmatrix} 5 & -5 \\ -15 & 15 \end{bmatrix},$$

and hence the corresponding global stiffness calculated via the assembly of the elemental matrices is

$$\begin{bmatrix} 20 & -5 & & & & & & & 0 \\ -15 & 20 & -5 & & & & & & 0 \\ 0 & -15 & 20 & -5 & & & & & 0 \\ 0 & 0 & -15 & 20 & -5 & & & & 0 \\ 0 & \cdots & & -15 & 20 & -5 & & \cdots & 0 \\ 0 & & \cdots & & -15 & 20 & -5 & 0 & 0 \\ 0 & & & \cdots & & -15 & 20 & -5 & 0 \\ 0 & & & & \cdots & & -15 & 20 & -5 \\ 0 & & & & & \cdots & & -15 & 20 \end{bmatrix}.$$

As expected the elemental and global matrices corresponding to the CD equation are not symmetric. Using the above calculated matrix the standard Galerkin solution is carried out. As shown in Fig. 3.14 the standard Galerkin method with the adopted level of discretization is capable of producing a stable-accurate solution, therefore, in this case the CD equation does not represent a multiscale behavior.

Example 12. CD equation — exact RFB, $P_e = 20$ and $l = 0.1$.

In this example the Peclet number is increased to demonstrate the effects of the onset of multiscale behavior on the standard Galerkin and bubble function-based methods.

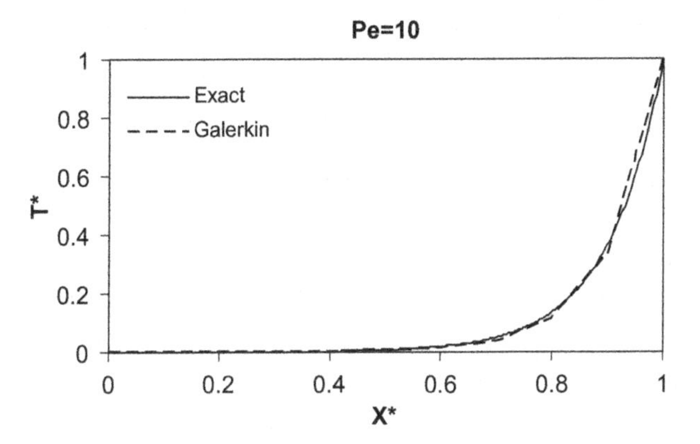

Fig. 3.14 Comparison of the standard Galerkin and analytical solution ($P_e = 10$ and $l = 0.1$).

After the insertion of given data, the ESM corresponding to the standard Galerkin method in this case is calculated as

$$\text{ESM} = \begin{bmatrix} 0 & 0 \\ -20 & 20 \end{bmatrix},$$

and the global stiffness matrix is calculated as

$$\begin{bmatrix} 20 & 0 & \dots & & & & & & 0 \\ -20 & 20 & 0 & \dots & & & & & 0 \\ 0 & -20 & 20 & 0 & \dots & & & & 0 \\ 0 & 0 & -20 & 20 & 0 & \dots & & & 0 \\ 0 & 0 & 0 & -20 & 20 & 0 & \dots & & 0 \\ 0 & \dots & & & -20 & 20 & 0 & 0 & 0 \\ 0 & & \dots & & & -20 & 20 & 0 & 0 \\ 0 & & & \dots & & & -20 & 20 & 0 \\ 0 & & & & & & & -20 & 20 \end{bmatrix}.$$

The ESM derived using bubble shape functions in their exact analytical form (given by Eq. 3.58) is

$$\text{ESM} = \begin{bmatrix} \dfrac{-P_e}{1 - e^{P_e x}} & \dfrac{P_e}{1 - e^{P_e x}} \\ \dfrac{P_e e^{P_e l}}{1 - e^{P_e x}} & \dfrac{-P_e e^{P_e l}}{1 - e^{P_e x}} \end{bmatrix}. \tag{3.65}$$

Insertion of the numerical values for the Peclet number and e, we have

$$\text{ESM} = \begin{bmatrix} 3.1304 & -3.1304 \\ -23.1304 & 23.1304 \end{bmatrix}.$$

Therefore, after the assembly of the elemental matrices the global stiffness matrix in this case is found as

$$\begin{bmatrix}
26.2607 & -3.1304 & & & & & & & \\
-23.1304 & 26.2607 & -3.1304 & & & & & & \text{\Huge 0} \\
0 & -23.1304 & 26.2607 & -3.1304 & & & & & \\
 & & -23.1304 & 26.2607 & -3.1304 & & & & \\
 & & & -23.1304 & 26.2607 & -3.1304 & & & \\
 & & & & -23.1304 & 26.2607 & -3.1304 & & \\
 & & & & & -23.1304 & 26.2607 & -3.1304 & 0 \\
\text{\Huge 0} & & & & & & -23.1304 & 26.2607 & -3.1304 \\
 & & & & & & & -23.1304 & 26.2607
\end{bmatrix}.$$

Using the described data, solutions corresponding to the standard Galerkin and Bubble function schemes are obtained. Comparison of the results obtained by the standard Galerkin and exact RFB schemes with the analytical solution is shown in Fig. 3.15. As is evident from Fig. 3.15, standard Galerkin scheme fails to generate an accurate solution.

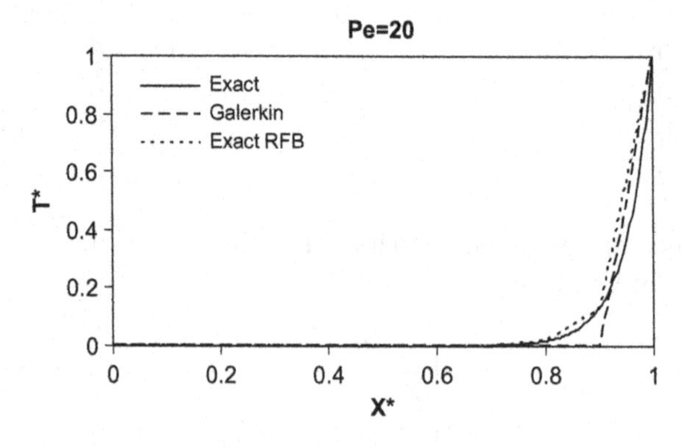

Fig. 3.15 Comparison of the results generated by the standard Galerkin, analytical, and exact RFB methods ($P_e = 20$ and $l = 0.1$).

Table 3.10. Comparison of numerical values generated by the exact RFB method with analytical solution at nodal points ($P_e = 20$ and $l = 0.1$).

X	Exact	Exact RFB
0.0	0	0
0.1	0	0
0.2	0	0
0.3	0	0
0.4	0	0
0.5	0	0
0.6	0.0003	0.0003
0.7	0.0025	0.0025
0.8	0.0183	0.0183
0.9	0.1353	0.1353
1.0	1	1

In Table 3.10, the perfect match between the nodal values generated by the exact RFB scheme and the analytical solution for this case are shown.

Example 13. CD equation — exact RFB, $P_e = 100$ and $l = 0.1$.

We now consider a case in which the multiscale nature of governing CD equation is stronger. To this end we choose $P_e = 100$ and $l = 0.1$. In this case the ESM corresponding to the exact RFB is found as

$$\text{ESM} = \begin{bmatrix} 0.0045 & -0.0045 \\ -100.0045 & 100.0045 \end{bmatrix},$$

and the global stiffness matrix is found after the assembly of the elemental matrices as

$$\begin{bmatrix}
100.0091 & -0.0045 & & & & & & & \\
-100.0045 & 100.091 & -0.0045 & & & & & \text{\Huge 0} & \\
0 & -100.0045 & 100.091 & -0.0045 & & & & & \\
 & & -100.0045 & 100.091 & -0.0045 & & & & \\
 & & & -100.0045 & 100.091 & -0.0045 & & & \\
 & & & & -100.0045 & 100.091 & -0.0045 & & \\
 & & & & & -100.0045 & 100.091 & -0.0045 & 0 \\
 & \text{\Huge 0} & & & & & -100.0045 & 100.091 & -0.0045 \\
 & & & & & & & -100.0045 & 100.091
\end{bmatrix}.$$

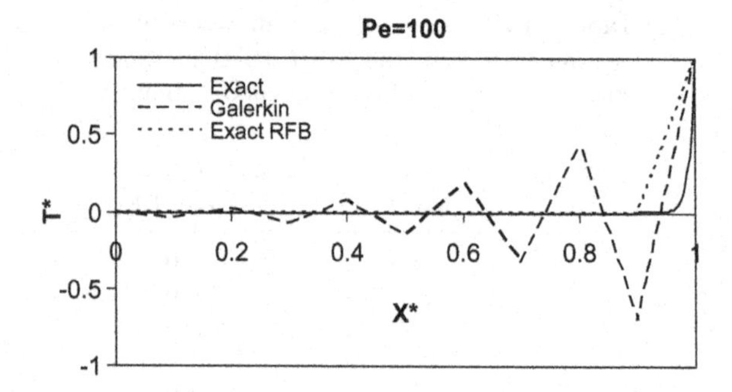

Fig. 3.16 Comparison of the results generated by the standard Galerkin, analytical and exact RFB methods ($P_e = 100$ and $l = 0.1$).

Table 3.11. Comparison of numerical results generated by the exact RFB method with analytical solution at nodal points ($P_e = 100$ and $l = 0.1$).

X	Exact	Exact RFB
0.0	0	0
0.1	0	0
0.2	0	0
0.3	0	0
0.4	0	0
0.5	0	0
0.6	0	0
0.7	0	0
0.8	0	0
0.9	0	0
1.0	1	1

Numerical results obtained using both the exact RFB and standard Galerkin schemes are compared with the analytical solution in Fig. 3.16. As shown in this figure, the standard Galerkin scheme fails to generate stable results in this case. In contrast, as shown in Table 3.11, the exact RFB scheme generates exact solutions at the nodal points.

3.3.3 *Multiscale finite element solution of the CDR equation using RFB function method*

To consider a wider range of physical conditions that give rise to multiscale behavior in reactive transport processes, we consider the solution of

CDR equation. This equation is solved using different stabilized methods (Hoffman *et al.*, 2005; Burman and Hansbo, 2004; Alhumaizi, 2003; Tezduyar and Park, 1986).

Steady state CDR equation given in a dimensionless form is written as

$$\left(\frac{dT^*}{dx^*}\right) - \frac{1}{P_e}\left(\frac{d^2T^*}{dx^{*2}}\right) - D_aT^* = f^*, \tag{3.66}$$

where P_e is Peclet number and D_a is Damköhler number (note that the definition of Damköhler number in DR and CDR equations can be different, see Chap. 4). Similar to previous cases to obtain a node-based solution for Eq. (3.66) in a domain discretized into two-noded linear elements, the field variable in CDR equation T^* should be replaced in terms of its nodal values and shape functions N_1 and N_2. Therefore, nodal one-dimensional forms of Eq. (3.66) are written as

$$\begin{cases} -\dfrac{1}{P_e}\dfrac{d^2N_1}{dx^2} + \dfrac{dN_1}{dx} - D_aN_1 = 0 \quad \text{for } x \in [0-l] \\ \begin{cases} N_1(0) = 1 \\ N_1(l) = 0 \end{cases} \end{cases} \tag{3.67}$$

and

$$\begin{cases} -\dfrac{1}{P_e}\dfrac{d^2N_2}{dx^2} + \dfrac{dN_2}{dx} - D_aN_2 = 0 \quad \text{for } x \in [0-l] \\ \begin{cases} N_2(0) = 0 \\ N_2(l) = 1 \end{cases} \end{cases}, \tag{3.68}$$

where l is a characteristic element length and x is elemental coordinate variable ($x = 0 - 1$). Assuming conditions under which $P_e^2 - 4P_eD_a > 0$, the solution of Eqs. (3.67) and (3.68) gives the analytical shape functions which, using a local elemental coordinate system, are expressed as

$$\begin{cases} N_1 = \dfrac{\exp((0.5P_e - \alpha)l + (0.5P_e + \alpha)x) - \exp((0.5P_e + \alpha)l + (0.5P_e - \alpha)x)}{\exp(0.5P_e - \alpha)l - \exp(0.5P_e + \alpha)l} \\ N_2 = \dfrac{\exp(0.5P_e - \alpha)x - \exp(0.5P_e + \alpha)x}{\exp(0.5P_e - \alpha)l - \exp(0.5P_e + \alpha)l} \end{cases}, \tag{3.69}$$

where

$$\alpha = 0.5\sqrt{P_e^2 - 4P_eD_a}. \tag{3.70}$$

After the substitution of the exponential function by truncated Taylor series (in a manner similar to the one represented by Eq. (3.60)) in Eq. (3.69) truncated forms of analytical shape functions corresponding to Eqs. (3.67) and (3.68) are found as

$$N_1 = \frac{(A_2 - A_1)\,x + (A_1 - A_2)\,l + \frac{1}{2!}\left[\left(A_2^2 - A_1^2\right)x^2 + \left(A_1^2 - A_2^2\right)l^2\right]}{(A_1 - A_2)\,l + \frac{1}{2!}\left(A_1^2 - A_2^2\right)l^2}, \tag{3.71}$$

where

$$\begin{cases} A_1 = 0.5P_e + \alpha \\ A_2 = 0.5P_e - \alpha \end{cases}. \tag{3.72}$$

Rearranging

$$N_1 = \frac{l - x}{l} + \frac{b_2}{(a + b_2 l)\,l}x\,(l - x), \tag{3.73}$$

where

$$a = (A_1 - A_2) \tag{3.74}$$

and

$$b_2 = \frac{1}{2!}\left(A_1^2 - A_2^2\right). \tag{3.75}$$

Similarly for N_2 we have

$$N_2 = \frac{(A_2 - A_1)\,x + \frac{1}{2!}\left(A_2^2 - A_1^2\right)x^2}{(A_1 - A_2)\,l + \frac{1}{2!}\left(A_1^2 - A_2^2\right)l^2}. \tag{3.76}$$

Rearranging

$$N_2 = \frac{x}{l} + \frac{b_2}{(a + b_2 l)\,l}x\,(l - x). \tag{3.77}$$

The first term in Eqs. (3.73) and (3.77) corresponds to normal linear shape functions for a two-node element and the second terms represent the bubble functions. Using the described procedure the third-order bubble-enriched (RFB) shape functions for the CDR equation are found as

$$\begin{cases} N_1 = \frac{l - x}{l} + \frac{b_1 + b_2 + 2lb_3}{(a + b_1 l + b_3 l^2)\,l}x(l - x) + \frac{b_3}{(a + b_1 l + b_3 l^2)\,l}(l - x)x(l - x) \\ N_2 = \frac{x}{l} + \frac{b_1 + b_3 l}{(a + b_1 l + b_3 l^2)\,l}x(l - x) + \frac{b_3}{(a + b_1 l + b_3 l^2)\,l}xx(l - x) \end{cases}, \tag{3.78}$$

where

$$a = A_1 - A_2, \quad b_1 = \frac{1}{2!}\left(A_1^2 - A_2^2\right),$$

$$b_2 = \frac{3l}{3!}\left(A_2 A_1^2 - A_1 A_2^2\right), \quad b_3 = \frac{1}{3!}\left(A_1^3 - A_2^3\right).$$

Similarly the fourth-order bubble-enriched (RFB) shape functions are found as

$$\begin{cases} N_1 = \dfrac{l-x}{l} + \dfrac{b_1 + b_2 + 2lb_3 + 2lb_4 + 3l^2 b_5}{(a + b_1 l + b_4 l^2 + b_5 l^3)\, l} x(l-x) \\ \qquad - \dfrac{b_3 + b_4 + 2lb_5}{(a + b_1 l + b_4 l^2 + b_5 l^3)\, l}(l-x)x(l-x) \\ \qquad - \dfrac{b_5}{(a + b_1 l + b_4 l^2 + b_5 l^3)\, l} x^2 (l-x)^2 \\ N_2 = \dfrac{x}{l} - \dfrac{b_2 + b_4 l + b_5 l^2}{(a + b_1 l + b_4 l^2 + b_5 l^3)\, l} x(l-x) \\ \qquad - \dfrac{b_4 + 2b_5 l}{(a + b_1 l + b_4 l^2 + b_5 l^3)\, l} x x(l-x) \\ \qquad + \dfrac{b_5}{(a + b_1 l + b_4 l^2 + b_5 l^3)\, l} x^2 (l-x)^2 \end{cases}, \tag{3.79}$$

where

$$a = A_1 - A_2, \quad b_1 = \frac{l}{2}\left(A_2 A_1^2 - A_1 A_2^2\right), \quad b_2 = \frac{1}{2}\left(A_1^2 - A_2^2\right),$$

$$b_3 = \frac{l}{3!}\left(A_2 A_1^3 - A_1 A_2^3\right), \quad b_4 = \frac{1}{3!}\left(A_1^3 - A_2^3\right), \quad b_5 = \frac{1}{4!}\left(A_1^4 - A_2^4\right).$$

The fifth-order bubble-enriched (RFB) shape functions are found as

$$\begin{cases} N_1 = \dfrac{l-x}{l} + \dfrac{b_1 + b_2 + 2lb_3 + lb_4 + 2lb_5 + 3l^2 b_6 + 3l^2 b_7 + 4l^3 b_8}{C} \\ \qquad \times x(l-x) - \dfrac{b_3 + b_4 + b_5 + 2lb_6 + 2lb_7 + 3l^2 b_8}{C}(l-x)x(l-x) \\ \qquad - \dfrac{b_6 + b_7 + 3lb_8}{C} x^2 (l-x)^2 + \dfrac{b_8}{C}(l-x)\, x^2 (l-x)^2 \\ N_2 = \dfrac{x}{l} - \dfrac{b_2 + b_5 l + b_7 l^2 + b_8 l^3}{C} x(l-x) - \dfrac{b_5 + 2b_7 l + 3l^2 b_8}{C} x x(l-x) \\ \qquad + \dfrac{b_7 + 2lb_8}{C} x^2 (l-x)^2 + \dfrac{b_8}{C} x x^2 (l-x)^2 \end{cases}, \tag{3.80}$$

where

$$C = \left(a + b_2 l + b_5 l^2 + b_7 l^3 + b_8 l^4\right) l \tag{3.81}$$

and

$$a = A_1 - A_2, \quad b_1 = \frac{l}{2}\left(A_2 A_1^2 - A_1 A_2^2\right), \quad b_2 = \frac{1}{2}\left(A_1^2 - A_2^2\right),$$

$$b_3 = \frac{l}{3!}\left(A_2 A_1^3 - A_1 A_2^3\right), \quad b_4 = \frac{10 l^2}{5!}\left(A_2^2 A_1^3 - A_2^3 A_1^2\right),$$

$$b_5 = \frac{1}{3!}\left(A_1^3 - A_2^3\right), \quad b_6 = \frac{5l}{5!}\left(A_2 A_1^4 - A_1 A_2^4\right),$$

$$b_7 = \frac{1}{4!}\left(A_1^4 - A_2^4\right), \quad b_8 = \frac{1}{5!}\left(A_1^5 - A_2^5\right).$$

The described process can be repeated to find other higher-order bubble function-enriched shape functions. Using the derived shape functions the ESM corresponding to the given CDR equation is found as

$$\begin{bmatrix} \displaystyle\int_0^\ell \left(\frac{d\psi_1}{dx}\frac{dw_1}{dx} + P_e w_1 \frac{dN_1}{dx} - P_e D_a w_1 N_1\right) dx & \displaystyle\int_0^\ell \left(\frac{d\psi_2}{dx}\frac{dw_1}{dx} + P_e w_1 \frac{dN_2}{dx} - P_e D_a w_1 N_2\right) dx \\ \displaystyle\int_0^\ell \left(\frac{d\psi_1}{dx}\frac{dw_2}{dx} + P_e w_2 \frac{dN_1}{dx} - P_e D_a w_2 N_1\right) dx & \displaystyle\int_0^\ell \left(\frac{d\psi_2}{dx}\frac{dw_2}{dx} + P_e w_2 \frac{dN_2}{dx} - P_e D_a w_2 N_2\right) dx \end{bmatrix}. \tag{3.82}$$

However, in the finite element procedure the weight functions are taken to be identical to the ordinary linear shape functions

$$w_1 = \frac{l - x}{l}, \quad w_2 = \frac{x}{l}. \tag{3.83}$$

Therefore the ESM is calculated as

$$\text{ESM} = \begin{bmatrix} E_{11} & E_{12} \\ E_{21} & E_{22} \end{bmatrix}, \tag{3.84}$$

where

$$E_{11} = \frac{1}{l} - P_e - \frac{P_e}{l}\frac{\frac{1}{A_1}e^{A_2 l}(1-e^{A_1 l}) - \frac{1}{A_2}e^{A_1 l}(1-e^{A_2 l})}{e^{A_2 l} - e^{A_1 l}}$$

$$- P_e D_a \frac{\frac{1}{A_2}e^{A_1 l} - \frac{1}{A_1}e^{A_2 l}}{e^{A_2 l} - e^{A_1 l}} + \frac{P_e D_a}{l}$$

$$\times \frac{\frac{1}{A_1^2}e^{A_2 l}(1-e^{A_1 l}) - \frac{1}{A_2^2}e^{A_1 l}(1-e^{A_2 l})}{e^{A_2 l} - e^{A_1 l}},$$

$$E_{12} = \frac{-1}{l} - \frac{P_e}{l}\frac{\frac{1}{A_1}(1-e^{A_1 l}) - \frac{1}{A_2}(1-e^{A_2 l})}{e^{A_1 l} - e^{A_2 l}} - P_e D_a \frac{\frac{1}{A_2} - \frac{1}{A_1}}{e^{A_1 l} - e^{A_2 l}}$$

$$+ \frac{P_e D_a}{l}\frac{\frac{1}{A_1^2}(1-e^{A_1 l}) - \frac{1}{A_2^2}(1-e^{A_2 l})}{e^{A_1 l} - e^{A_2 l}},$$

$$E_{21} = \frac{-1}{l} + \frac{P_e}{l}\frac{\frac{1}{A_1}e^{A_2 l}(1-e^{A_1 l}) - \frac{1}{A_2}e^{A_1 l}(1-e^{A_2 l})}{e^{A_2 l} - e^{A_1 l}} - \frac{P_e D_a}{l}$$

$$\times \frac{\frac{1}{A_1^2}e^{A_2 l}(1-e^{A_1 l}) - \frac{1}{A_2^2}e^{A_1 l}(1-e^{A_2 l}) + l(\frac{1}{A_1}e^{A_1 l + A_2 l} - \frac{1}{A_2}e^{A_1 l + A_2 l})}{e^{A_2 l} - e^{A_1 l}},$$

$$E_{22} = \frac{1}{l} + P_e + \frac{P_e}{l}\frac{\frac{1}{A_1}(1-e^{A_1 l}) - \frac{1}{A_2}(1-e^{A_2 l})}{e^{A_1 l} - e^{A_2 l}}$$

$$- \frac{P_e D_a}{l}\frac{\frac{1}{A_1^2}(1-e^{A_1 l}) - \frac{1}{A_2^2}(1-e^{A_2 l}) + l(\frac{1}{A_1}e^{A_1 l} - \frac{1}{A_2}e^{A_2 l})}{e^{A_1 l} - e^{A_2 l}}.$$

Example 14. CDR equation — exact RFB, $P_e = 10$, $D_a = -20$, and $l = 0.1$.

We now consider the solution of the following problem:

$$\begin{cases} \left(\frac{dT^*}{dx^*}\right) - \frac{1}{P_e}\left(\frac{d^2T^*}{dx^{*2}}\right) - D_a T^* = 0 \\ \text{Subject to} \\ T^* = 0 \quad \text{at } x^* = 0 \\ T^* = 1 \quad \text{at } x^* = 1 \end{cases}.$$

The ESM corresponding to the exact RFB method in this case is found as

$$\text{ESM} = \begin{bmatrix} 11.5719 & -4.2728 \\ 11.6146 & 21.5719 \end{bmatrix},$$

and after the assembly of the elemental matrices

$$
\begin{bmatrix}
33.1437 & -4.2728 & & & & & & & \\
-11.6146 & 33.1437 & -4.2728 & & & & & \mathbf{0} & \\
0 & -11.6146 & 33.1437 & -4.2728 & & & & & \\
 & & -11.6146 & 33.1437 & -4.2728 & & & & \\
 & & & -11.6146 & 33.1437 & -4.2728 & & & \\
 & & & & -11.6146 & 33.1437 & -4.2728 & & \\
 & & & & & -11.6146 & 33.1437 & -4.2728 & 0 \\
\mathbf{0} & & & & & & -11.6146 & 33.1437 & -4.2728 \\
 & & & & & & & -11.6146 & 33.1437
\end{bmatrix}.
$$

Nodal solutions obtained using the exact RFB scheme are compared with the analytical solution in Table 3.12 which shows a perfect match. The standard Galerkin finite element solution of the present problem fails to generate a stable solution with a discretization based on 10 equal size linear elements (i.e. the mesh used here).

Example 15. CDR equation-exact RFB, $P_e = 100$, $D_a = -20$, and $l = 0.1$.

Table 3.12. Comparison of numerical values generated by the exact RFB method with analytical solution at nodal points ($P_e = 10$, $D_a = -20$, and $l = 0.1$).

X	Exact	Exact RFB
0.0	0	0
0.1	0	0
0.2	0	0
0.3	0	0
0.4	0	0
0.5	0	0
0.6	0.0003	0.0003
0.7	0.0025	0.0025
0.8	0.0183	0.0183
0.9	0.1363	0.1353
1.0	1	1

To examine an extreme case, we repeat the above solution this time using an increased Peclet number of 100 (i.e. the process is convection dominated). Results obtained using the standard Galerkin based on ordinary elements, analytical method, and the exact RFB scheme are shown in Fig. 3.17. As expected, the standard Galerkin scheme fails to generate a stable solution.

In contrast, as shown in Table 3.13, the exact RFB scheme maintains stability and generates accurate results at nodal points.

Example 16. CDR equation — second-order RFB, $P_e = 10$, $D_a = -20$, and $l = 0.1$.

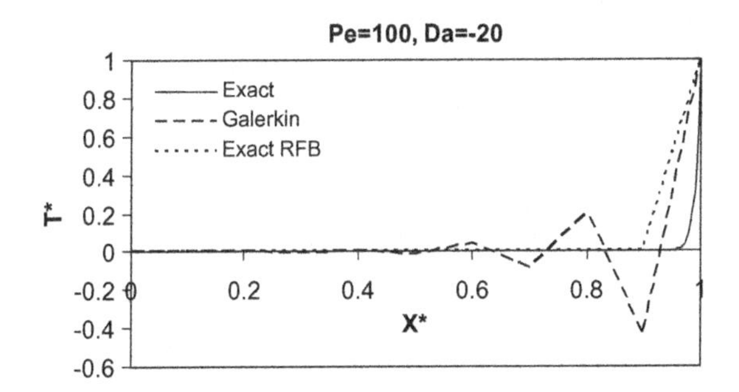

Fig. 3.17 Comparison of results generated by standard Galerkin, analytical, and RFB methods ($P_e = 100$, $D_a = -20$, and $l = 0.1$).

Table 3.13. Comparison of nodal values generated by the exact RFB method with analytical solution ($P_e = 100, D_a = -20$, and $l = 0.1$).

X	Exact	Exact RFB
0.0	0	0
0.1	0	0
0.2	0	0
0.3	0	0
0.4	0	0
0.5	0	0
0.6	0	0
0.7	0	0
0.8	0	0
0.9	0	0
1.0	1	1

As can be expected, the degree of accuracy of the results generated using the described bubble function enrichments depends on the degree of the bubble function used to enrich the shape functions of the element employed to solve a problem. However, even using a very low-order bubble function the scheme should generate results that are more accurate than the standard Galerkin finite element scheme based on ordinary elements. To demonstrate this point we compare the solution of homogeneous CDR equation using the standard Galerkin and second-order RFB schemes for a low Peclet number dissipative reaction case. The terms of ESM corresponding to the second-order RFB method are found from Eqs. (3.73) and (3.77) as

$$
\begin{aligned}
E_{11} &= \int_0^l \left\{ \left(\frac{-1}{l}\right)\left(\frac{-1}{l}\right) + P_e\left(\frac{l-x}{l}\right)\left(\frac{-1}{l} + b(l-2x)\right) \right. \\
&\quad \left. - P_e D_a \left(\frac{l-x}{l}\right)\left(\frac{l-x}{l} + bx(l-x)\right)\right\} dx, \\
E_{12} &= \int_0^l \left\{ \left(\frac{1}{l}\right)\left(\frac{-1}{l}\right) + P_e\left(\frac{l-x}{l}\right)\left(\frac{1}{l} + b(l-2x)\right) \right. \\
&\quad \left. - P_e D_a \left(\frac{l-x}{l}\right)\left(\frac{x}{l} + bx\,(l-x)\right)\right\} dx, \\
E_{21} &= \int_0^l \left\{ \left(\frac{-1}{l}\right)\left(\frac{1}{l}\right) + P_e\left(\frac{x}{l}\right)\left(\frac{-1}{l} + b(l-2x)\right) \right. \\
&\quad \left. - P_e D_a \left(\frac{x}{l}\right)\left(\frac{l-x}{l} + bx\,(l-x)\right)\right\} dx, \\
E_{22} &= \int_0^l \left\{ \left(\frac{1}{l}\right)\left(\frac{1}{l}\right) + P_e\left(\frac{x}{l}\right)\left(\frac{1}{l} + b(l-2x)\right) \right. \\
&\quad \left. - P_e D_a \left(\frac{x}{l}\right)\left(\frac{x}{l} + bx(l-x)\right)\right\} dx,
\end{aligned}
\tag{3.85}
$$

where

$$
b = \frac{b_2}{(a + b_2 l)\, l}, \tag{3.86}
$$

and $a = (A_1 - A_2)$, $b_2 = \frac{1}{2!}(A_1^2 - A_2^2)$, $\begin{cases} A_1 = 0.5P_e + \alpha \\ A_2 = 0.5P_e - \alpha \end{cases}$, taking $\alpha = 0.5\sqrt{P_e^2 - 4P_e D_a}$ then

$$b = \frac{b_2}{(a + b_2 l)\, l} = \frac{P_e \alpha}{(2\alpha + P_e \alpha l)\, l}.$$

Using the numerical values given in this example, b is calculated as $b = 50$. Note that in a local natural coordinate system of $\xi(-1, +1)$ we can replace x with $\frac{l}{2}(1 + \xi)$ in the set of Eq. (3.85) to obtain $b\frac{l^2}{4} = 0.125$. The bubble coefficient is the same at both nodes of a linear element. Using these values the ESM is calculated as

$$\mathrm{ESM} = \begin{bmatrix} 11.6667 & -3.3333 \\ -11.6667 & 20.0000 \end{bmatrix},$$

and the corresponding global matrix is

$$\begin{bmatrix}
31.6667 & -3.3333 & & & & & & & \\
-11.6667 & 31.6667 & -3.3333 & & & & & \mathbf{0} & \\
0 & -11.6667 & 31.6667 & -3.3333 & & & & & \\
 & & -11.6667 & 31.6667 & -3.3333 & & & & \\
 & & & -11.6667 & 31.6667 & -3.3333 & & & \\
 & & & & -11.6667 & 31.6667 & -3.3333 & & \\
 & & & & & -11.6667 & 31.6667 & -3.3333 & 0 \\
 & \mathbf{0} & & & & & -11.6667 & 31.6667 & -3.3333 \\
 & & & & & & & -11.6667 & 31.6667
\end{bmatrix}.$$

Solutions obtained via the standard Galerkin, second-order RFB schemes and the analytical result for the problem posed in this example are shown in Fig. 3.18. The second-order RFB scheme has generated noticeably better results than the standard Galerkin scheme. Both results can be improved if a more refined mesh is used. However, in the case of bubble-enriched technique, apart from mesh refinement, higher-order enrichments can also be readily utilized to obtain better results.

Example 17. CDR equation — exact RFB, $P_e = 10$, $D_a = 20$, and $l = 0.1$.

To investigate the performance of the RFB method under various physical conditions, we consider a propagation case represented by the insertion of $P_e = 10$ and $D_a = 20$ into the CDR equation. The ESM derived using

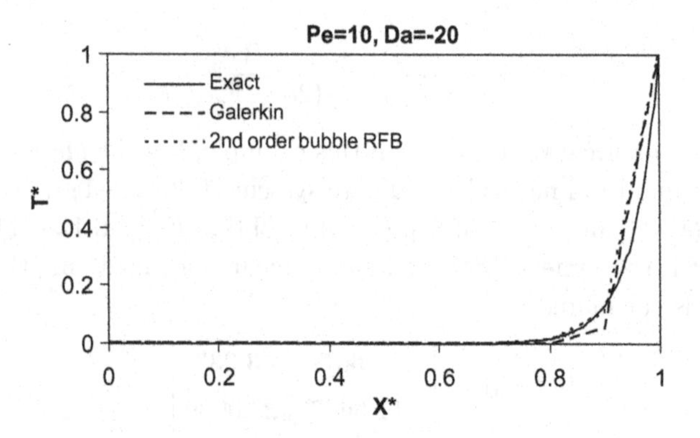

Fig. 3.18 Comparison of the results generated by the standard Galerkin, second-order RFB function and analytical methods ($P_e = 10$, $D_a = -20$, and $l = 0.1$).

the standard Galerkin method is

$$\text{ESM} = \begin{bmatrix} -1.6667 & -8.3333 \\ -18.3333 & 8.3333 \end{bmatrix},$$

and the corresponding global matrix is found as

$$\begin{bmatrix} 6.6667 & -8.3333 & & & & & & & \\ -18.3333 & 6.6667 & -8.3333 & & & & & \text{\huge 0} & \\ 0 & -18.3333 & 6.6667 & -8.3333 & & & & & \\ & & -18.3333 & 6.6667 & -8.3333 & & & & \\ & & & -18.3333 & 6.6667 & -8.3333 & & & \\ & & & & -18.3333 & 6.6667 & -8.3333 & & \\ & & & & & -18.3333 & 6.6667 & -8.3333 & 0 \\ \text{\huge 0} & & & & & & -18.3333 & 6.6667 & -8.3333 \\ & & & & & & & -18.3333 & 6.6667 \end{bmatrix}.$$

The elemental and global matrices corresponding to the exact RFB method are found as

$$\text{ESM} = \begin{bmatrix} -1.6514 & -8.2767 \\ -22.4984 & 8.3486 \end{bmatrix},$$

and

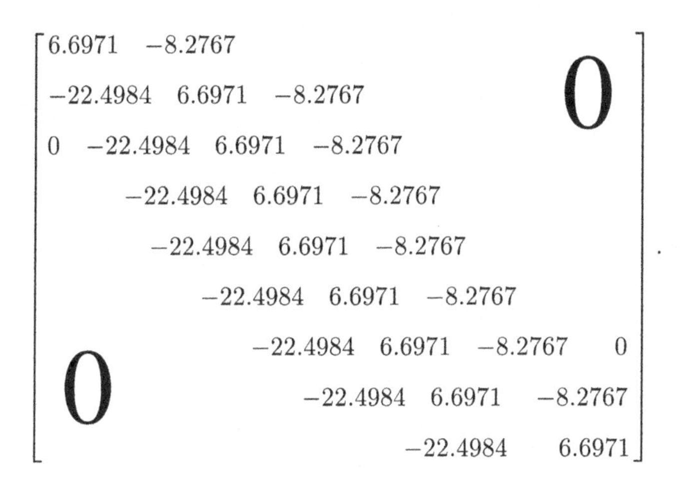

$$\begin{bmatrix} 6.6971 & -8.2767 & & & & & & & \\ -22.4984 & 6.6971 & -8.2767 & & & & & \mathbf{0} & \\ 0 & -22.4984 & 6.6971 & -8.2767 & & & & & \\ & & -22.4984 & 6.6971 & -8.2767 & & & & \\ & & & -22.4984 & 6.6971 & -8.2767 & & & \\ & & & & -22.4984 & 6.6971 & -8.2767 & & \\ & & & & & -22.4984 & 6.6971 & -8.2767 & 0 \\ & \mathbf{0} & & & & & -22.4984 & 6.6971 & -8.2767 \\ & & & & & & & -22.4984 & 6.6971 \end{bmatrix}.$$

Solutions generated by these methods are shown in Fig. 3.19 and compared with the analytical results.

As shown in Table 3.14 the exact RFB results are identical to the analytical solution at the nodes.

Example 18. CDR equation — exact RFB, $P_e = 10$, $D_a = 200$, and $l = 0.1$.

We repeat the solution for an extreme propagation situation, and again, as shown in Fig. 3.20 and Table 3.15, the exact RFB method generates significantly superior results to those obtained by the standard Galerkin scheme based on ordinary elements.

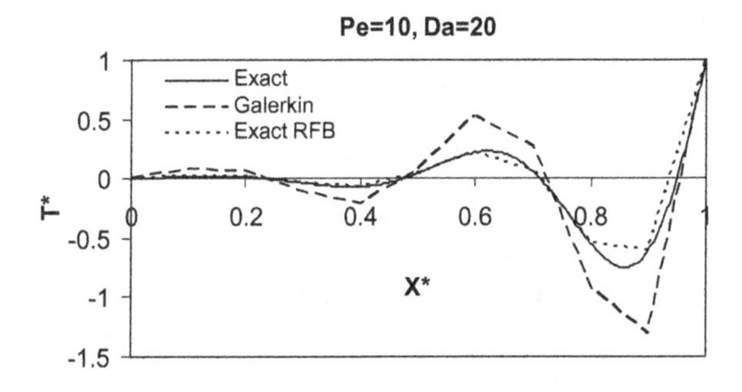

Fig. 3.19 Results generated by the standard Galerkin, exact RFB function and analytical methods ($P_e = 10$, $D_a = 20$, and $l = 0.1$).

Table 3.14. Comparison of nodal values generated by the exact RFB method with analytical solution (P_e = 10, D_a = 20, and $l = 0.1$).

X	Exact	Exact RFB
0.0	0	0
0.1	0.0175	0.0175
0.2	0.0142	0.0142
0.3	−0.0361	−0.0361
0.4	−0.0678	−0.0678
0.5	0.0434	0.0434
0.6	0.2193	0.2193
0.7	0.0595	0.0595
0.8	−0.548	−0.548
0.9	−0.6051	−0.6051
1.0	1	1

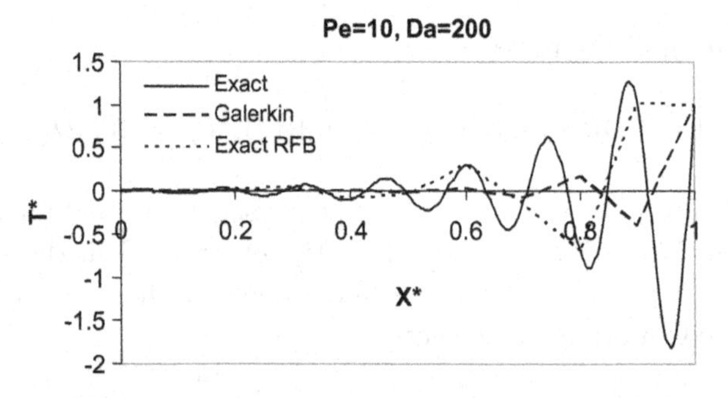

Fig. 3.20 Results generated by the standard Galerkin, exact RFB function, and analytical methods (P_e = 10, D_a = 200, and $l = 0.1$).

3.4 Practical Implementation of the STC Method

In previous sections it is explained that the derivation of the RFB schemes depends on the availability of an analytical solution for the original governing equation of a multiscale problem. However, under most types of realistic conditions it is often impossible to obtain such a solution. This difficulty is avoided using a method called the Static Condensation (STC). Using this technique, the governing differential equations representing a problem can be used to incorporate any desired bubble function with ordinary

Table 3.15. Comparison of nodal values generated by the exact RFB method with analytical solution ($P_e = 10$, $D_a = 200$, and $l = 0.1$).

X	Exact	Exact RFB
0.0	0	0
0.1	−0.0242	−0.0242
0.2	0.0211	0.0211
0.3	0.0473	0.0473
0.4	−0.0988	−0.0988
0.5	−0.0421	−0.0421
0.6	0.3054	0.3054
0.7	−0.1524	−0.1524
0.8	−0.697	−0.697
0.9	1.0236	1.0236
1.0	1	1

Lagrangian shape functions. The basic concept of the STC method is briefly explained earlier in this chapter and fundamental methods used in its formulation are given by Eqs. (3.15) and (3.16). In this section, the main steps of the STC method are further described and procedures used to derive practical schemes are illustrated via solved examples.

Historically, incorporation of the STC-based bubble functions in the standard Galerkin finite element schemes was proposed before the development of variational multiscale frameworks by Arnold *et al.* (1984) and Baiocchi *et al.* (1993).

The main steps of the STC technique can be summarized as shown below.

Step 1. Select the desired bubble function (any function which is zero at the element boundaries) to be incorporated with the ordinary Lagrangian shape functions.

Step 2. Using weighted residual statement of the problem apply the STC technique using Eqs. (3.15) and (3.16).

Step 3. Continue with the normal weighted residual finite element procedures as detailed in Chap. 1.

We now consider the solution of a number of typical multiscale transport problems using the STC method.

3.4.1 *Multiscale finite element solution of the DR equation using STC method*

In principle, any desired bubble function can be incorporated with linear shape functions using the STC method. Note that in the RFB method a process which is equivalent to the "condensation" automatically takes place during the derivation of bubble functions and bubble coefficients are calculated as a part of the condensation procedure. We now consider a series of polynomial functions which have the property of vanishing on the nodes of a linear element represented by elemental coordinate system of $-1 \leq \xi \leq +1$

$$\phi_b = (1 - \xi^2). \tag{3.87}$$

In this context we call the function represented by Eq. (3.87) a second-order bubble function. Similarly a fourth-order elemental bubble function is defined as

$$\phi_b = (1 - \xi^2) + (1 - \xi^2)^2 \tag{3.88}$$

and a sixth-order bubble function is given as

$$\phi_b = (1 - \xi^2) + (1 - \xi^2)^2 + (1 - \xi^2)^3. \tag{3.89}$$

Accordingly the nth-order elemental bubble function may be written as

$$\phi_b = (1 - \xi^2) + (1 - \xi^2)^2 + (1 - \xi^2)^3 + \cdots + (1 - \xi^2)^n = \sum_{q=1}^{n} (1 - \xi^2)^q. \tag{3.90}$$

We now describe the steps required to develop a solution for the homogeneous DR equation, written in a dimensionless form as

$$\frac{\partial^2 T^*}{\partial x^{*2}} - D_a T^* = 0. \tag{3.91}$$

Step 1. Bubble functions are incorporated with ordinary linear Lagrangian shape functions to obtain

$$\tilde{T} = \psi_1 T_1 + \psi_2 T_2 + \phi_b T_3, \tag{3.92}$$

where ψ_i is the Lagrangian linear shape function and ϕ_b represents the selected bubble function.

Step 2. The above approximation is used to formulate the required weighted residual statement of Eq. (3.91) as

$$\int_0^l \left(\left(\frac{d\psi_1}{dx}\frac{dw_b}{dx} + D_a w_b \psi_1 \right) T_1 + \left(\frac{d\psi_2}{dx}\frac{dw_b}{dx} + D_a w_b \psi_2 \right) T_2 \right.$$
$$\left. + \left(\frac{d\phi_b}{dx}\frac{dw_b}{dx} + D_a w_b \phi_b \right) T_3 \right) dx = 0, \tag{3.93}$$

where $w_b = \phi_b$, therefore,

$$T_3 = \frac{\int_0^\ell \left(\frac{d\psi_1}{dx}\frac{dw_b}{dx} + D_a w_b \psi_1 \right) dx}{\int_0^\ell \left(\frac{d\phi_b}{dx}\frac{dw_b}{dx} + D_a w_b \phi_b \right) dx} T_1 + \frac{\int_0^\ell \left(\frac{d\psi_2}{dx}\frac{dw_b}{dx} + D_a w_b \psi_2 \right) dx}{\int_0^\ell \left(\frac{d\phi_b}{dx}\frac{dw_b}{dx} + D_a w_b \phi_b \right) dx} T_2. \tag{3.94}$$

Replacing T_3 in Eq. (3.92), we have

$$\tilde{T} = \left(\psi_1 + \frac{\int_0^\ell \left(\frac{d\psi_1}{dx}\frac{dw_b}{dx} + D_a w_b \psi_1 \right) dx}{\int_0^\ell \left(\frac{d\phi_b}{dx}\frac{dw_b}{dx} + D_a w_b \phi_b \right) dx} \phi_b \right) T_1$$
$$+ \left(\psi_2 + \frac{\int_0^\ell \left(\frac{d\psi_2}{dx}\frac{dw_b}{dx} + D_a w_b \psi_2 \right) dx}{\int_0^\ell \left(\frac{d\phi_b}{dx}\frac{dw_b}{dx} + D_a w_b \phi_b \right) dx} \phi_b \right) T_2. \tag{3.95}$$

In a multiscale approach the shape functions are taken to be a combination of ordinary and bubble functions as $N_i = \psi_i + b_i \phi_b$.

Therefore using a linear Lagrangian element (which is the most convenient and natural choice), we have

$$\begin{cases} b_1 = \dfrac{\int_0^\ell \left(\frac{d\psi_1}{dx}\frac{dw_b}{dx} + D_a w_b \psi_1 \right) dx}{\int_0^\ell \left(\frac{d\phi_b}{dx}\frac{dw_b}{dx} + D_a w_b \phi_b \right) dx} \\[2ex] b_2 = \dfrac{\int_0^\ell \left(\frac{d\psi_2}{dx}\frac{dw_b}{dx} + D_a w_b \psi_2 \right) dx}{\int_0^\ell \left(\frac{d\phi_b}{dx}\frac{dw_b}{dx} + D_a w_b \phi_b \right) dx} \end{cases}. \tag{3.96}$$

Using the equations set of (3.96), the bubble coefficients corresponding to any selected bubble function can be evaluated. For example, for the second-order bubble function given by Eq. (3.87) $b_1 = b_2 = b$, we have

$$b = -\left(1.60 + \frac{16}{D_a l^2}\right)^{-1}. \tag{3.97}$$

Similarly for the fourth-, sixth-, and eighth-order bubble function we have

$$b = -\left(3.09 + \frac{31.238}{D_a l^2}\right)^{-1}, \tag{3.98}$$

$$b = -\left(4.52 + \frac{49.643}{D_a l^2}\right)^{-1}, \tag{3.99}$$

and

$$b = -\left(5.92 + \frac{71.554}{D_a l^2}\right)^{-1}, \tag{3.100}$$

respectively.

Step 3. ESM corresponding to the DR equation being solved here is now derived as

$$\text{ESM} = \begin{bmatrix} \int_0^\ell \left(\frac{d\psi_1}{dx}\frac{dw_1}{dx} + D_a w_1 N_1\right) dx & \int_0^\ell \left(\frac{d\psi_2}{dx}\frac{dw_1}{dx} + D_a w_1 N_2\right) dx \\ \int_0^\ell \left(\frac{d\psi_1}{dx}\frac{dw_2}{dx} + D_a w_2 N_1\right) dx & \int_0^\ell \left(\frac{d\psi_2}{dx}\frac{dw_2}{dx} + D_a w_2 N_2\right) dx \end{bmatrix}, \tag{3.101}$$

where N_i $(i = 1, 2)$ represents the enriched shape functions and ψ_i and w_i are ordinary linear shape and weight functions ($w_i = \psi_i$). Therefore, the ESM for the second-, fourth-, sixth-, and eighth-order STC-based bubble function schemes are

$$\begin{bmatrix} \frac{1}{l} + lD_a\left(\frac{1}{3} + b_1\frac{1}{3}\right) & \frac{-1}{l} + lD_a\left(\frac{1}{6} + b_2\frac{1}{3}\right) \\ -\frac{1}{l} + lD_a\left(\frac{1}{6} + b_1\frac{1}{3}\right) & \frac{1}{l} + lD_a\left(\frac{1}{3} + b_2\frac{1}{3}\right) \end{bmatrix}, \tag{3.102}$$

$$\begin{bmatrix} \frac{1}{l} + lD_a\left(\frac{1}{3} + b_1\frac{3}{5}\right) & \frac{-1}{l} + lD_a\left(\frac{1}{6} + b_2\frac{3}{5}\right) \\ -\frac{1}{l} + lD_a\left(\frac{1}{6} + b_1\frac{3}{5}\right) & \frac{1}{l} + lD_a\left(\frac{1}{3} + b_2\frac{3}{5}\right) \end{bmatrix}, \tag{3.103}$$

$$\begin{bmatrix} \frac{1}{l} + lD_a\left(\frac{1}{3} + 0.8286b_1\right) & \frac{-1}{l} + lD_a\left(\frac{1}{6} + 0.8286b_2\right) \\ -\frac{1}{l} + lD_a\left(\frac{1}{6} + 0.8286b_1\right) & \frac{1}{l} + lD_a\left(\frac{1}{3} + 0.8286b_2\right) \end{bmatrix}, \tag{3.104}$$

and

$$\begin{bmatrix} \frac{1}{l} + lD_a\left(\frac{1}{3} + 1.0318b_1\right) & \frac{-1}{l} + lD_a\left(\frac{1}{6} + 1.0318b_2\right) \\ -\frac{1}{l} + lD_a\left(\frac{1}{6} + 1.0318b_1\right) & \frac{1}{l} + lD_a\left(\frac{1}{3} + 1.0318b_2\right) \end{bmatrix}, \tag{3.105}$$

respectively.

We now consider a number of solved examples where in each case the results generated by the STC, standard Galerkin, and analytical methods are compared.

Example 19. DR equation — second-order bubble function STC, $D_a = 500$, and $l = 0.1$.

After the insertion of the numerical values into the elemental matrix given by Eq. (3.102), we have

$$\text{ESM} = \begin{bmatrix} 23.2000 & -5.1333 \\ -5.1333 & 23.2000 \end{bmatrix}.$$

The corresponding global stiffness matrix is

$$\begin{bmatrix} 23.20 & -5.13 & & & & & & & & & \\ -5.13 & 46.40 & -5.13 & & & & & & & \mathbf{0} & \\ 0 & -5.13 & 46.40 & -5.13 & & & & & & & \\ & & -5.13 & 46.40 & -5.13 & & & & & & \\ & & & -5.13 & 46.40 & -5.13 & & & & & \\ & & & & -5.13 & 46.40 & -5.13 & & & & \\ & & & & & -5.13 & 46.40 & -5.13 & & & \\ & & & & & & -5.13 & 46.40 & -5.13 & & \\ & & & & & & & -5.13 & 46.40 & -5.13 & 0 \\ & \mathbf{0} & & & & & & & -5.13 & 46.40 & -5.13 \\ & & & & & & & & & -5.13 & 23.20 \end{bmatrix}.$$

Table 3.16. Comparison of nodal values generated by the standard Galerkin, second-order bubble function STC, and analytical methods (D_a = 500 and $l = 0.1$).

X	Exact	Standard Galerkin	Second-order STC
0.0	0	0	0
0.1	0	0	0
0.2	0	0	0
0.3	0	0	0
0.4	0	0	0
0.5	0	0	0
0.6	0.0001	0	0.0002
0.7	0.0012	0	0.0014
0.8	0.0114	0.001	0.0125
0.9	0.1069	0.0313	0.112
1.0	1	1	1

As shown in Table 3.16, the STC method (despite using a low-order bubble function) has generated a significantly more accurate solution than the standard Galerkin scheme based on ordinary elements.

Example 20. DR equation — second-order bubble function STC, D_a = 500, $l = 0.1$, and $f^* = 100$.

We now consider the solution of a nonhomogeneous DR equation using $D_a = 500$ and $f^* = 100$ in a mesh consisting of 10 elements of equal length (i.e. $l = 0.1$). Using the STC method, second-order bubble function in this case is found to be $\phi_f = f^* \left(\frac{8}{l^2} + 0.8D_a\right)^{-1} \phi_b$, where ϕ_b is the second-order bubble function given by Eq. (3.87). The ESM for this case is found using Eq. (3.52) as

$$\text{ESM} = \begin{bmatrix} 23.2 & -5.1333 \\ -5.1333 & 23.2 \end{bmatrix},$$

and the corresponding global matrix is found as

$$\begin{bmatrix} 46.4 & -5.1333 & 0 & 0 & 0 & 0 & 0 & 0 & 0 \\ -5.1333 & 46.4 & -5.1333 & 0 & 0 & 0 & 0 & 0 & 0 \\ 0 & -5.1333 & 46.4 & -5.1333 & 0 & 0 & 0 & 0 & 0 \\ 0 & 0 & -5.1333 & 46.4 & -5.1333 & 0 & 0 & 0 & 0 \\ 0 & 0 & 0 & -5.1333 & 46.4 & -5.1333 & 0 & 0 & 0 \\ 0 & 0 & 0 & 0 & -5.1333 & 46.4 & -5.1333 & 0 & 0 \\ 0 & 0 & 0 & 0 & 0 & -5.1333 & 46.4 & -5.1333 & 0 \\ 0 & 0 & 0 & 0 & 0 & 0 & -5.1333 & 46.4 & -5.1333 \\ 0 & 0 & 0 & 0 & 0 & 0 & 0 & -5.1333 & 46.4 \end{bmatrix}.$$

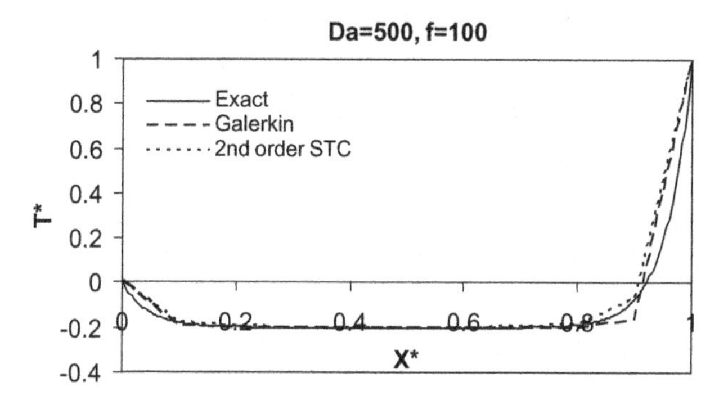

Fig. 3.21 Comparison of the results generated by the standard Galerkin, second-order bubble function STC and analytical methods ($D_a = 500$, $l = 0.1$, and $f^* = 100$).

The global load vector is also found using Eq. (3.52) as

$$\begin{bmatrix} -72.225 \\ -72.225 \\ -72.225 \\ -72.225 \\ -72.225 \\ -72.225 \\ -72.225 \\ -72.225 \\ -67.0917 \end{bmatrix}.$$

Using these data, solutions based on the standard Galerkin, second-order bubble function STC are obtained and compared with the analytical results (Fig. 3.21). As this figure shows, the STC scheme has generated a significantly more accurate result than the standard Galerkin scheme which uses ordinary elements.

3.4.2 *Multiscale finite element solution of the CD equation using STC method*

We now describe the steps required to develop a solution for the homogeneous CD equation, written in a dimensionless form as

$$\left(P_e \frac{\partial T^*}{\partial x^*}\right) - \left(\frac{\partial^2 T^*}{\partial x^{*2}}\right) = 0. \tag{3.106}$$

Step 1. As usual the STC scheme starts with the incorporation of a bubble function with ordinary linear Lagrangian shape functions as

$$\tilde{T} = \psi_1 T_1 + \psi_2 T_2 + \phi_b T_3, \tag{3.107}$$

where ψ_i is the linear Lagrangian shape function and ϕ_b represents the selected bubble function.

Step 2. Approximation shown in Eq. (3.107) is used to develop the weighted residual statement of the problem as

$$\int_0^\ell \left(\left(\frac{d\psi_1}{dx}\frac{dw_b}{dx} + P_e w_b \frac{d\psi_1}{dx} \right) T_1 + \left(\frac{d\psi_2}{dx}\frac{dw_b}{dx} + P_e w_b \frac{d\psi_2}{dx} \right) T_2 + \left(\frac{d\phi_b}{dx}\frac{dw_b}{dx} + P_e w_b \frac{d\phi_b}{dx} \right) T_3 \right) dx = 0, \tag{3.108}$$

where $w_b = \phi_b$ and therefore

$$T_3 = \frac{\int_0^\ell \left(\frac{d\psi_1}{dx}\frac{dw_b}{dx} + P_e w_b \frac{d\psi_1}{dx} \right) dx}{\int_0^\ell \left(\frac{d\phi_b}{dx}\frac{dw_b}{dx} + P_e w_b \frac{d\phi_b}{dx} \right) dx} T_1 + \frac{\int_0^\ell \left(\frac{d\psi_2}{dx}\frac{dw_b}{dx} + P_e w_b \frac{d\psi_2}{dx} \right) dx}{\int_0^\ell \left(\frac{d\phi_b}{dx}\frac{dw_b}{dx} + P_e w_b \frac{d\phi_b}{dx} \right) dx} T_2. \tag{3.109}$$

Substitution of T_3 from Eq. (3.109) into Eq. (3.107) gives

$$\tilde{T} = \left(\psi_1 + \frac{\int_0^\ell \left(\frac{d\psi_1}{dx}\frac{dw_b}{dx} + P_e w_b \frac{d\psi_1}{dx} \right) dx}{\int_0^\ell \left(\frac{d\phi_b}{dx}\frac{dw_b}{dx} + P_e w_b \frac{d\phi_b}{dx} \right) dx} \phi_b \right) T_1 + \left(\psi_2 + \frac{\int_0^\ell \left(\frac{d\psi_2}{dx}\frac{dw_b}{dx} + P_e w_b \frac{d\psi_2}{dx} \right) dx}{\int_0^\ell \left(\frac{d\phi_b}{dx}\frac{dw_b}{dx} + P_e w_b \frac{d\phi_b}{dx} \right) dx} \phi_b \right) T_2. \tag{3.110}$$

Similar to the solution of the DR equation, in a multiscale approach, the shape functions are taken to be a combination of ordinary and bubble functions as $N_i = \psi_i + b_i \phi_b$. Therefore, using a linear Lagrangian element (which is the most convenient and natural choice), we have

$$\begin{cases} b_1 = \dfrac{\int_0^\ell \left(\frac{d\psi_1}{dx}\frac{dw_b}{dx} + P_e w_b \frac{d\psi_1}{dx} \right) dx}{\int_0^\ell \left(\frac{d\phi_b}{dx}\frac{dw_b}{dx} + P_e w_b \frac{d\phi_b}{dx} \right) dx} \\ \\ b_2 = \dfrac{\int_0^\ell \left(\frac{d\psi_2}{dx}\frac{dw_b}{dx} + P_e w_b \frac{d\psi_2}{dx} \right) dx}{\int_0^\ell \left(\frac{d\phi_b}{dx}\frac{dw_b}{dx} + P_e w_b \frac{d\phi_b}{dx} \right) dx} \end{cases}. \tag{3.111}$$

Before evaluation of the bubble coefficients for selected bubble functions, the relationships shown in the equation set of (3.111) are transformed into a local natural coordinate system of $\xi(-1, +1)$ by replacing x with $\frac{l}{2}(1+\xi)$.

Therefore,

$$\begin{cases} b_1 = \dfrac{\int_{-1}^{+1}\left(\frac{2}{l}\frac{d\psi_1}{d\xi}\frac{dw_b}{d\xi} + P_e w_b \frac{d\psi_1}{d\xi}\right)d\xi}{\int_{-1}^{+1}\left(\frac{2}{l}\frac{d\phi_b}{d\xi}\frac{dw_b}{d\xi} + P_e w_b \frac{d\phi_b}{d\xi}\right)d\xi} \\ b_2 = \dfrac{\int_{-1}^{+1}\left(\frac{2}{l}\frac{d\psi_2}{d\xi}\frac{dw_b}{d\xi} + P_e w_b \frac{d\psi_2}{d\xi}\right)d\xi}{\int_{-1}^{+1}\left(\frac{2}{l}\frac{d\phi_b}{d\xi}\frac{dw_b}{d\xi} + P_e w_b \frac{d\phi_b}{d\xi}\right)d\xi} \end{cases}. \tag{3.112}$$

Using these relationships the bubble coefficient corresponding to bubble functions expressed as

$$\phi_b = (1 - \xi^{2n}) \tag{3.113}$$

are found to be

$$\begin{cases} b_1 = \dfrac{n(4n-1)}{8n^2(2n+1)} l P_e \quad \text{and} \\ b_2 = -b_1 \end{cases}. \tag{3.114}$$

for the bubble functions of the form

$$\begin{aligned} \phi_b &= (1-\xi^2) + (1-\xi^2)^2 + (1-\xi^2)^3 + \cdots + (1-\xi^2)^n \\ &= \sum_{q=1}^{n} (1-\xi^2)^q. \end{aligned} \tag{3.115}$$

The coefficient for second- and fourth-order bubble functions are

$$\begin{aligned} b_1 &= \frac{1}{8} l P_e, \\ b_2 &= -b_1, \end{aligned}$$

and

$$\begin{aligned} b_1 &= \frac{1}{15.6191} l P_e, \\ b_2 &= -b_1. \end{aligned}$$

Step 3. The general form of the ESM for the CD equation is now found as

$$\begin{bmatrix} \int_0^l \left(\dfrac{d\psi_1}{dx}\dfrac{dw_1}{dx} + P_e w_1 \dfrac{dN_1}{dx}\right) dx & \int_0^l \left(\dfrac{d\psi_2}{dx}\dfrac{dw_1}{dx} + P_e w_1 \dfrac{dN_2}{dx}\right) dx \\ \int_0^l \left(\dfrac{d\psi_1}{dx}\dfrac{dw_2}{dx} + P_e w_2 \dfrac{dN_1}{dx}\right) dx & \int_0^l \left(\dfrac{d\psi_2}{dx}\dfrac{dw_2}{dx} + P_e w_2 \dfrac{dN_2}{dx}\right) dx \end{bmatrix}. \tag{3.116}$$

This yields the following matrix for the second-order bubble function

$$\mathrm{ESM} = \begin{bmatrix} \left(-\frac{1}{2}+\frac{2}{3}b_1\right)P_e+\frac{1}{l} & \left(\frac{1}{2}+\frac{2}{3}b_2\right)P_e-\frac{1}{l} \\ \left(-\frac{1}{2}+\frac{2}{3}b_1\right)P_e-\frac{1}{l} & \left(\frac{1}{2}+\frac{2}{3}b_2\right)P_e+\frac{1}{l} \end{bmatrix}, \tag{3.117}$$

and for the fourth-order bubble function we obtain

$$\mathrm{ESM} = \begin{bmatrix} \left(-\frac{1}{2}+\frac{6}{5}b_1\right)P_e+\frac{1}{l} & \left(\frac{1}{2}+\frac{6}{5}b_2\right)P_e-\frac{1}{l} \\ \left(-\frac{1}{2}+\frac{6}{5}b_1\right)P_e-\frac{1}{l} & \left(\frac{1}{2}+\frac{6}{5}b_2\right)P_e+\frac{1}{l} \end{bmatrix}. \tag{3.118}$$

The derived ESM is now used to solve the following example.

Example 21. CD equation — polynomial STC bubble function, $P_e = 20$ and $l = 0.1$

In this case, using the given numerical values, the bubble coefficients for the second-order STC method are found as

$$b_1 = \frac{1}{8}lP_e = \frac{2}{8} = 0.25,$$
$$b_2 = -b_1 = -0.25.$$

Hence, the ESM for the second-order bubble function is

$$\mathrm{ESM} = \begin{bmatrix} 3.3333 & -3.3333 \\ -16.6667 & 16.6667 \end{bmatrix}.$$

The corresponding global stiffness matrix is

$$\begin{bmatrix} 20 & -3.3333 & & & & & & & \\ -16.6667 & 20 & -3.3333 & & & & & \mathbf{0} & \\ 0 & -16.6667 & 20 & -3.3333 & & & & & \\ & & -16.6667 & 20 & -3.3333 & & & & \\ & & & -16.6667 & 20 & -3.3333 & & & \\ & & & & -16.6667 & 20 & -3.3333 & & \\ & & & & & -16.6667 & 20 & -3.3333 & 0 \\ & \mathbf{0} & & & & & -16.6667 & 20 & -3.3333 \\ & & & & & & & -16.6667 & 20 \end{bmatrix}.$$

In Fig. 3.22, the solutions generated by the standard Galerkin and the second-order bubble function STC methods are presented and compared

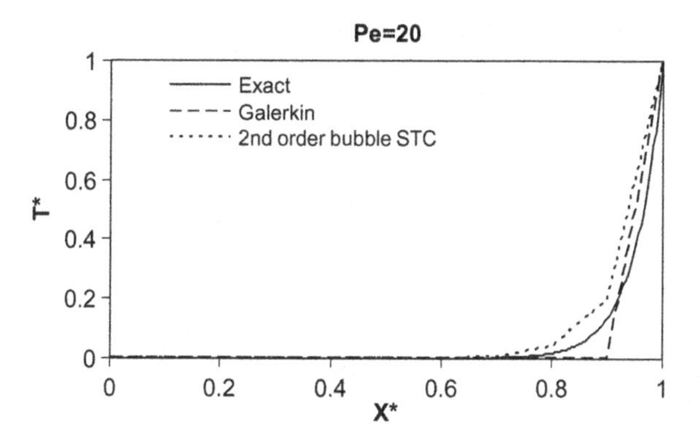

Fig. 3.22 Comparison of the results generated by the standard Galerkin, second-order bubble function STC and analytical methods ($P_e = 20$ and $l = 0.1$).

with the analytical result. Although both methods have produced stable results they seem to be inaccurate.

We now repeat the above solution this time using the fourth-order bubble function STC scheme. The bubble coefficients in this case are found as

$$\begin{aligned} b_1 &= \frac{1}{15.6191} l P_e = 0.128, \\ b_2 &= -b_1 = -0.128. \end{aligned}$$

The ESM for the fourth-order bubble function is

$$\mathrm{ESM} = \begin{bmatrix} 3.0720 & -3.0720 \\ -16.9280 & 16.9280 \end{bmatrix}.$$

The corresponding global stiffness matrix is

$$\begin{bmatrix} 20 & -3.0720 & & & & & & & & \\ -16.9280 & 20 & -3.0720 & & & & & & \mathbf{0} & \\ 0 & -16.9280 & 20 & -3.0720 & & & & & & \\ & & -16.9280 & 20 & -3.0720 & & & & & \\ & & & -16.9280 & 20 & -3.0720 & & & & \\ & & & & -16.9280 & 20 & -3.0720 & & & \\ & & & & & -16.9280 & 20 & -3.0720 & 0 & \\ & \mathbf{0} & & & & & -16.9280 & 20 & -3.0720 & \\ & & & & & & & -16.9280 & 20 & \end{bmatrix}.$$

Table 3.17. Comparison of the results generated by the Second- and Fourth-order STC methods with the standard Galerkin and analytical solutions at nodal points ($P_e = 20$ and $l = 0.1$).

X	Standard Galerkin	Second STC	Fourth STC	Exact
0.0	0	0	0	0
0.1	0	0	0	0
0.2	0	0	0	0
0.3	0	0	0	0
0.4	0	0.0001	0	0
0.5	0	0.0003	0.0002	0
0.6	0	0.0016	0.0011	0.0003
0.7	0	0.008	0.006	0.0025
0.8	0	0.04	0.0329	0.0183
0.9	0	0.2	0.1815	0.1353
1.0	1	1	1	1

The nodal solutions obtained in this case are given in Table 3.17 and compared with the results generated by the standard Glaerkin, second-order bubble function STC, and analytical methods. As shown in this table, the fourth-order method yields more accurate results.

3.4.3 *Multiscale finite element solution of the CDR equation using STC method*

To complete the discussions presented here we now consider the solution of CDR equation. Using a dimensionless form the CDR equation is written as

$$\left(\frac{dT^*}{dx^*}\right) - \frac{1}{P_e}\left(\frac{d^2T^*}{dx^{*2}}\right) - D_a T^* = 0. \tag{3.119}$$

Step 1. The solution procedure again starts with the representation of the field variable appearing in Eq. (3.119) in terms of bubble functions with ordinary linear Lagrangian shape functions as

$$\tilde{T} = \psi_1 T_1 + \psi_2 T_2 + \phi_b T_3, \tag{3.120}$$

where ψ_i is the linear Lagrangian shape function and ϕ_b represents a selected bubble function.

Step 2. The above approximation is used to construct the weighted residual statement of the problem as

$$\int_0^\ell \left(\left(\frac{d\psi_1}{dx}\frac{dw_b}{dx} + P_e w_b \frac{d\psi_1}{dx} - P_e D_a w_b \phi_b \right) T_1 \right.$$
$$+ \left(\frac{d\psi_2}{dx}\frac{dw_b}{dx} + P_e w_b \frac{d\psi_2}{dx} - P_e D_a w_b \phi_b \right) T_2$$
$$\left. + \left(\frac{d\phi_b}{dx}\frac{dw_b}{dx} + P_e w_b \frac{d\phi_b}{dx} - P_e D_a w_b \phi_b \right) T_3 \right) dx, \quad (3.121)$$

where $w_b = \phi_b$ and therefore

$$T_3 = \left(\psi_1 + \frac{\int_0^\ell \left(\frac{d\psi_1}{dx}\frac{dw_b}{dx} + P_e w_b \frac{d\psi_1}{dx} - P_e D_a w_b \psi_1 \right) dx}{\int_0^\ell \left(\frac{d\phi_b}{dx}\frac{dw_b}{dx} + P_e w_b \frac{d\phi_b}{dx} - P_e D_a w_b \phi_b \right) dx} \phi_b \right) T_1$$
$$+ \left(\psi_2 + \frac{\int_0^\ell \left(\frac{d\psi_2}{dx}\frac{dw_b}{dx} + P_e w_b \frac{d\psi_2}{dx} - P_e D_a w_b \psi_2 \right) dx}{\int_0^\ell \left(\frac{d\phi_b}{dx}\frac{dw_b}{dx} + P_e w_b \frac{d\phi_b}{dx} - P_e D_a w_b \phi_b \right) dx} \phi_b \right) T_2. \quad (3.122)$$

Repeating the procedure described in the solution of DR and CD equations, we have

$$b_1 = \frac{\int_0^\ell \left(\frac{d\psi_1}{dx}\frac{dw_b}{dx} + P_e w_b \frac{d\psi_1}{dx} - P_e D_a w_b \psi_1 \right) dx}{\int_0^\ell \left(\frac{d\phi_b}{dx}\frac{dw_b}{dx} + P_e w_b \frac{d\phi_b}{dx} - P_e D_a w_b \phi_b \right) dx},$$
$$b_2 = \frac{\int_0^\ell \left(\frac{d\psi_2}{dx}\frac{dw_b}{dx} + P_e w_b \frac{d\psi_2}{dx} - P_e D_a w_b \psi_2 \right) dx}{\int_0^\ell \left(\frac{d\phi_b}{dx}\frac{dw_b}{dx} + P_e w_b \frac{d\phi_b}{dx} - P_e D_a w_b \phi_b \right) dx}. \quad (3.123)$$

Step 3. The general ESM in this case is derived as

$$\begin{bmatrix} \int_0^\ell \left(\frac{d\psi_1}{dx}\frac{dw_1}{dx} + P_e w_1 \frac{dN_1}{dx} - P_e D_a w_1 N_1 \right) dx & \int_0^\ell \left(\frac{d\psi_2}{dx}\frac{dw_1}{dx} + P_e w_1 \frac{dN_2}{dx} - P_e D_a w_1 N_2 \right) dx \\ \int_0^\ell \left(\frac{d\psi_1}{dx}\frac{dw_2}{dx} + P_e w_2 \frac{dN_1}{dx} - P_e D_a w_2 N_1 \right) dx & \int_0^\ell \left(\frac{d\psi_2}{dx}\frac{dw_2}{dx} + P_e w_2 \frac{dN_2}{dx} - P_e D_a w_2 N_2 \right) dx \end{bmatrix}. \quad (3.124)$$

For the following type of bubble functions

$$\phi_b = (1 - \xi^{2n}), \quad n = 1, 2, 3, 4, \ldots \quad (3.125)$$

The enriched shape functions are written as

$$\begin{cases} N_1 = \frac{1}{2}(1-\xi) + b_1(1-\xi^{2n}) \\ N_2 = \frac{1}{2}(1+\xi) + b_2(1-\xi^{2n}) \end{cases}. \tag{3.126}$$

Therefore, the bubble coefficients b_1 and b_2 are calculated as

$$\begin{cases} b_1 = \left[\frac{l}{2}D_a\left(1-\frac{1}{2n+1}\right) - \left(\frac{1}{2n+1}-1\right)\right]\left(\frac{1}{\beta}\right) \\ b_2 = \left[\frac{l}{2}D_a\left(1-\frac{1}{2n+1}\right) + \left(\frac{1}{2n+1}-1\right)\right]\left(\frac{1}{\beta}\right) \\ \beta = \frac{16n^2}{lP_e}\frac{1}{4n-1} - \frac{l}{2}D_a\left(\frac{2}{2n+1}-\frac{2}{4n^2+1}\right) \end{cases}, \tag{3.127}$$

where l is element length. For a second-order bubble function ($n = 1$)

$$\begin{cases} b_1 = \left(\frac{lD_a}{3}+\frac{2}{3}\right)\frac{1}{\beta} \\ b_2 = \left(\frac{lD_a}{3}-\frac{2}{3}\right)\frac{1}{\beta} \\ \beta = \frac{16}{3lP_e} - lD_a\frac{2}{15} \end{cases}, \tag{3.128}$$

and the general form of the ESM for second-order bubble is hence found as

$$\begin{bmatrix} \left(-\frac{1}{2}-\frac{2}{3}b_1\right)P_e + \frac{1}{l} & \left(\frac{1}{2}+\frac{2}{3}b_2\right)P_e - \frac{1}{l} \\ -lP_eDa\left(\frac{1}{3}+b_1\frac{1}{3}\right) & -lP_eD_a\left(\frac{1}{6}+b_2\frac{1}{3}\right) \\ \left(-\frac{1}{2}-\frac{2}{3}b_1\right)P_e - \frac{1}{l} & \left(\frac{1}{2}+\frac{2}{3}b_1\right)P_e + \frac{1}{l} \\ -lP_eDa\left(\frac{1}{6}+b_1\frac{1}{3}\right) & -lP_eDa\left(\frac{1}{3}+b_2\frac{1}{3}\right) \end{bmatrix}. \tag{3.129}$$

The derived general form is now used to solve the following example.

Example 22. CDR equation — second-order bubble function STC $P_e = 10$, $D_a = -10$, and $l = 0.1$

In this case the insertion of the numerical data into the general forms gives

$$\begin{cases} b_1 = \dfrac{15}{82} = 0.183 \\[2ex] b_2 = -\dfrac{5}{82} = -0.061 \end{cases}.$$

The elemental and global matrices corresponding to the standard Galerkin method in this case are as follows

$$\text{ESM} = \begin{bmatrix} 8.3333 & -3.3333 \\ -13.3333 & 18.3333 \end{bmatrix}$$

and

$$\begin{bmatrix}
26.6667 & -3.3333 & & & & & & & \\
-13.3333 & 26.6667 & -3.3333 & & & & & & \mathbf{0} \\
0 & -13.3333 & 26.6667 & -3.3333 & & & & & \\
 & & -13.3333 & 26.6667 & -3.3333 & & & & \\
 & & & -13.3333 & 26.6667 & -3.3333 & & & \\
 & & & & -13.3333 & 26.6667 & -3.3333 & & \\
 & & & & & -13.3333 & 26.6667 & -3.3333 & 0 \\
\mathbf{0} & & & & & & -13.3333 & 26.6667 & -3.3333 \\
 & & & & & & & -13.3333 & 26.6667
\end{bmatrix}.$$

Table 3.18. Comparison of the results generated by the second-order STC methods with the standard Galerkin and analytical solutions at nodal points ($P_e = 10$, $Da = -10$, and $l = 0.1$).

X	Exact	Standard Galerkin	Second STC
0.0	0	0	0
0.1	0	0	0
0.2	0	0	0
0.3	0	0	0
0.4	0.0001	0	0
0.5	0.0003	0	0.0001
0.6	0.0016	0.0003	0.0009
0.7	0.0077	0.0024	0.005
0.8	0.0389	0.0179	0.0292
0.9	0.1972	0.134	0.171
1.0	1	1	1

The elemental and global stiffness matrices representing the second-order bubble function STC scheme are

$$\text{ESM} = \begin{bmatrix} 7.7233 & -3.9433 \\ -13.9433 & 17.7233 \end{bmatrix}$$

and

$$\begin{bmatrix}
25.4467 & -3.9433 & & & & & & & \\
-13.9433 & 25.6667 & -3.9433 & & & & & & \mathbf{0} \\
0 & -13.9433 & 25.6667 & -3.9433 & & & & & \\
 & & -13.9433 & 25.6667 & -3.9433 & & & & \\
 & & & -13.9433 & 25.6667 & -3.9433 & & & \\
 & & & & -13.9433 & 25.6667 & -3.9433 & & \\
 & & & & & -13.9433 & 25.6667 & -3.9433 & 0 \\
\mathbf{0} & & & & & & -13.9433 & 25.6667 & -3.9433 \\
 & & & & & & & -13.9433 & 25.6667
\end{bmatrix}.$$

Solutions generated by the standard Galerkin and the second-order bubble function STC schemes are shown in Table 3.18 and compared with the analytical results. As shown in this table, the STC scheme's results are closer to the analytical solution.

References

[1] Alhumaizi, K., Henda, R. and Soliman, M., Numerical analysis of a reaction–diffusion–convection system, *Comput. Chem. Eng.*, 2003; 27; 579–594.

[2] Arnold, D.N., Brezzi, F. and Fortin, M., A stable finite element for the Stokes equations, *Calcolo*, 1984; 23; 337–344.

[3] Baiocchi, C., Brezzi, F. and Franca, L.P., Virtual bubbles and the Galerkin/least squares method, *Comput. Meth. Appl. Mech. Eng.*, 1993; 105; 125–142.

[4] Brezzi, F., Bristeau, M.O., Franca, L.P., Mallet, M. and Roge, G., A relationship between stabilized finite element methods and the Galerkin method with bubble functions, *Comput. Meth. Appl. Mech. Eng.*, 1992; 96; 117–129.

[5] Brezzi, F. and Russo, A., Choosing bubbles for advection-diffusion problems, *Math. Models Meth. Appl. Sci.*, 1994; 4; 571–587.

[6] Brezzi, F., Franca L.P., Hughes, T.J.R. and Russo, A., b =g. *Comput. Meth. Appl. Mech. Eng.*, 1997; 145; 329–339.

[7] Brezzi, F., Franca, L.P. and Russo, A., Further considerations on residual-free bubbles for advective-diffusive equations, *Comput. Meth. Appl. Mech. Eng.*, 1998; 166; 25–33.

[8] Brezzi, F. and Marini, L.D., Augmented spaces, two-level methods, and stabilizing subgrids, *Int. J. Numer. Meth. Fluids,* 2002; 40; 31–46.

[9] Brezzi, F., Hauke, G., Marini, L.D. and Sangalli, G., Link-cutting bubbles for convection–diffusion–reaction problems, *Math. Models Meth. Appl. Sci.,* 2003; 13; 445–461.

[10] Brezzi, F., Marini, L.D., and Russo, A., On the choice of a stabilizing subgrid for convection-diffusion problems, *Comput. Meth. Appl. Mech. Eng.*, 2005; 194; 127–148.

[11] Brooks, A. and Hughes, T.J.R., Streamline upwind/Petrov–Galerkin methods for advection dominated flows, *Proc. Third Int. Conf. Finite Element Methods in Fluid Flow*, Banff, Canada, 1980.

[12] Brooks, A.N. and Hughes, T.J.R., Streamline Upwind/Petrov–Galerkin formulations for convection dominated flows with particular emphasis on the incompressible Navier–Stokes equations, *Comput. Meth. Appl. Mech. Eng.*, 1982; 32; 199–259.

[13] Burman, E. and Hansbo, P., Edge stabilization for Galerkin approximations of convection–diffusion–reaction problems, *Comput. Meth. Appl. Mech. Eng.*, 2004; 193; 1437–1453.

[14] Christie, I., Griffiths, D.F., Mitchell A.R. and Zienkiewicz, O.C., Finite element methods for second order differential equations with significant first derivatives, *Int. J. Numer. Meth. Eng.*, 1976; 10; 1389–1396.

[15] Codina, R., Comparison of some finite element methods for solving the diffusion–convection–reaction equation, *Comput. Meth. Appl. Mech. Eng.*, 1998; 156; 185–210.

[16] Donea, J. and Huerta, A., *Finite Element Methods for Flow Problems*, John Wiley & Sons, Chichester, 2003.

[17] Farhat, C., Harari, I. and Franca, L.P., The discontinuous enrichment method, *Comput. Meth. Appl. Mech. Eng.*, 2001; 190; 6455–6479.

[18] Farhat, C., Harari, I. and Hetmaniuk U., A discontinuous Galerkin method with Lagrange multipliers for the solution of Helmholtz problems in the mid-frequency regime. *Comput. Meth. Appl. Mech. Eng.*, 2003a; 192; 1389–1419.

[19] Farhat, C., Harari, I. and Hetmaniuk, U., The discontinuous enrichment method for multiscale analysis, *Comput. Meth. Appl. Mech. Eng.*, 2003b; 192; 3195–3210.

[20] Franca, L.P. and Russo, A., Deriving upwinding, mass lumping and selective reduced integration by residual free bubbles, *Appl. Math. Lett.*, 1996; 9; 83–88.

[21] Franca, L.P., Hughes, T.J.R. and Stenberg, R., Stabilized finite element methods for the stokes problem, in *Incompressible Fluid Dynamics-Trends and Advances*, Nicolaides, R.A. and Gunzberger, M.D. (Eds.), Cambridge University Press, Cambridge, 1993.

[22] Franca, L.P. and Farhat, C., Bubble functions prompt unusual stabilized finite element methods, *Comput. Meth. Appl. Mech. Eng.*, 1995; 123; 299–308.

[23] Franca, L.P., Farhat, C., Macedo, A.P. and Lesoinne, M., Residual free bubbles for Helemholtz equation. *Comput. Meth. Appl. Mech. Eng.*, 1997; 40; 4003–4009.

[24] Franca, L.P. and Russo, A., Mass lumping emanating from residual free bubbles, *Comput. Meth. Appl. Mech. Eng.*, 1997a; 142; 353–360.

[25] Franca, L.P. and Russo, A., Unlocking with residual-free bubbles. *Comput. Meth. Appl. Mech. Eng.*, 1997a; 142; 361–364.

[26] Hauke, G., A simple subgrid scale stabilized method for the advection–diffusion–reaction equation, *Comput. Meth. Appl. Mech. Eng.*, 2002; 191; 2925–2947.

[27] Heinrich, J.C., Huyakorn, P.S., Zienkiewicz, O.C. and Mitchell, A.R., An upwind finite element scheme for two-dimensional convective transport equation, *Int. J. Numer. Meth. Eng.*, 1977; 11; 134–143.

[28] Hoffman, J., Johnson, C. and Bertoluzza, S., Subgrid modeling for convection–diffusion–reaction in one space dimension using a Haar Multiresolution analysis, *Comput. Meth. Appl. Mech. Eng.*, 2005; 194; 19–44.

[29] Hughes, T.J.R., A simple scheme for developing upwind finite elements, *Int. J. Numer. Meth. Eng.*, 1978; 12; 1359–1365.

[30] Hughes, T.J.R. and Brooks, A.N., A multidimensional upwind scheme with no crosswind diffusion, in *Finite Element Methods for Convection Dominated Flows*, T.J.R. Hughes (Ed.), AMD; vol. 34; ASME, New York, 1979.

[31] Hughes, T.J.R., Multi scale phenomena, Green's functions, the Dirichlet-to-Neumann formulation, subgrid scale models, bubbles and the origins of stabilized methods, *Comput. Meth. Appl. Mech. Eng.*, 1995; 127; 381–401.

[32] Hughes, T.J.R. and Stewart, J., A space time formulation for multiscale phenomena, *Comput. Meth. Appl. Mech. Eng.*, 1996; 74; 217–229.

[33] Hughes, T.J.R., Feijoo, G.R., Mazzei, L. and Quincy, J.B., The variational multiscale method — a paradigm for computational mechanics, *Comput. Meth. Appl. Mech. Eng.*, 1998; 166; 3–24.

[34] Juanes, R. and Patzek T.W., Multiscale-stabilized solutions to one-dimensional systems of conservation laws, *Comput. Meth. Appl. Mech. Eng.*, 2005; 194; 2781–2805.

[35] Parvazinia, M., Nassehi, V., Wakeman, R.J. and Ghoreishy, M.H.R., Finite element modeling of flow through a porous medium between two parallel plates using the Brinkman equation, *Transport Porous Media*, 2006a; 63; 71–90.

[36] Parvazinia, M., Nassehi, V. and Wakeman, R.J., Multi-scale finite element modeling of laminar steady flow through highly permeable porous media, *Chem. Eng. Sci.*, 2006b; 61; 586–596.

[37] Russo, A., Bubble stabilization of finite element methods for the linearized incompressible Navier–Stokes equations, *Comput. Meth. Appl. Mech. Eng.*, 1996a; 132; 335–343.

[38] Russo A., A posteriori error estimators via bubble functions, *Math. Models Meth. Appl. Sci.* 1996b; 6; 33–41.

[39] Tezduyar, T.E. and Park, Y.J., Discontinuity-capturing finite element formulations for nonlinear convection–diffusion–reaction equations, *Comput. Meth. Appl. Mech. Eng.*, 1986; 59; 307–325.

CHAPTER 4

Simulation of Multiscale Transport Phenomena in Multidimensional Domains

In Chap. 3 the basic concepts and methodology used to develop multiscale (residual free bubble (RFB) and static condensation (STC)) finite element schemes are explained. However, in order to apply the described techniques to the solution of realistic problems they should be generalized to cope with multidimensional situations. In this chapter, the extension of the variational multiscale method to two-dimensional cases is presented. The extended technique is, subsequently, used to solve a number of benchmark problems. Similar to Chap. 3, problems posed in terms of the governing equations of transport phenomena provide the context in which the present multiscale schemes are discussed. The last section of this chapter outlines the extension of the multiscale approach to transient cases.

As explained in Chap. 3, the exact RFB method depends on element-based analytical solution of the governing equations of a given problem. In general, this is not possible in multidimensional cases and hence only approximate RFB based on polynomial bubble functions can be used.

4.1 Two-dimensional Multiscale Finite Element Technique

Consider the solution of the following transport equation over a continuous domain Ω.

$$\frac{\partial T}{\partial x} + \frac{\partial T}{\partial y} - \frac{1}{P_e}\left(\frac{\partial^2 T}{\partial x^2} + \frac{\partial^2 T}{\partial y^2}\right) - D_a T = 0, \tag{4.1}$$

where P_e and D_a are Peclet and Damköhler numbers, respectively. After the substitution of the field variable with an approximate form, weighting of the generated residual and application of Green's theorem to the second-order derivatives in Eq. (4.1) the following statement is constructed:

$$\int_{\Omega}\left(w\left(\frac{\partial\tilde{T}}{\partial x}+\frac{\partial\tilde{T}}{\partial y}\right)-\frac{1}{P_e}\left(\frac{\partial\tilde{T}}{\partial x}\frac{\partial w}{\partial x}+\frac{\partial\tilde{T}}{\partial y}\frac{\partial w}{\partial y}\right)-D_a w\tilde{T}\right)d\Omega=0. \quad (4.2)$$

Note that in Eq. (4.2) the boundary integral terms, which appear after the application of Green's theorem to the second-order derivatives, are ignored as they are not relevant to the present discussion.

In a rectangular master element in a coordinate system designated as the (x, y) system, we have

$$\int_0^{l_x}\int_0^{l_y}\left(w\left(T_x+T_y\right)+\frac{1}{P_e}\left(T_x w_x+T_y w_y\right)-D_a wT\right)dydx=0, \quad (4.3)$$

where T_x and T_y represent the x and y derivatives of $\tilde{T}$, respectively (to maintain simplicity the over bar indicating approximate form of the field variable is dropped).

Equation (4.3) is transformed into normalized local coordinates (ξ, η) using the following relationships

$$\frac{d}{dx}(\Phi)=\frac{2}{l}\frac{d}{d\xi}(\Phi), \quad \frac{d}{dy}(\Phi)=\frac{2}{l}\frac{d}{d\eta}(\Phi),$$

$$dx=\frac{l}{2}d\xi, \quad dy=\frac{l}{2}d\eta. \quad (4.4)$$

Assuming that $l_x = l_y = l$ and following the procedures outlined in Chaps. 2 and 3, a set of second-order enriched shape functions for a four-node square element are constructed as the tensor products of linear functions in ξ and η directions $(-1 \le \xi, \eta \le +1)$ as

$$\begin{cases} N_1=\frac{1}{4}(1-\xi)(1-\eta)+b_1(1-\xi^2)(1-\eta^2) \\ N_2=\frac{1}{4}(1+\xi)(1-\eta)+b_2(1-\xi^2)(1-\eta^2) \\ N_3=\frac{1}{4}(1+\xi)(1+\eta)+b_3(1-\xi^2)(1-\eta^2). \\ N_4=\frac{1}{4}(1-\xi)(1+\eta)+b_4(1-\xi^2)(1-\eta^2) \end{cases} \quad (4.5)$$

Conjunctive use of these shape functions with the Galerkin method gives the members of the elemental stiffness matrix corresponding to Eq. (4.3).

Therefore,

$$\begin{cases} E_{11} = \displaystyle\int_{-1}^{+1}\int_{-1}^{+1}\left(\frac{l}{2}w_1\left(N_{1\xi}+N_{1\eta}\right)+\frac{1}{P_e}\left(\psi_{1\xi}w_{1\xi}+\psi_{1\eta}w_{1\eta}\right)\right. \\ \qquad \left. -\dfrac{l^2}{4}D_aN_1w_1\right)d\xi d\eta \\ E_{12} = \displaystyle\int_{-1}^{+1}\int_{-1}^{+1}\left(\frac{l}{2}w_1\left(N_{2\xi}+N_{2\eta}\right)+\frac{1}{P_e}\left(\psi_{2\xi}w_{1\xi}+\psi_{2\eta}w_{1\eta}\right)\right. \\ \qquad \left. -\dfrac{l^2}{4}D_aN_2w_1\right)d\xi d\eta \\ \vdots \text{ etc} \end{cases} \quad (4.6)$$

After the integration and substitution of D_aP_e by D

$$E_{11} = \frac{l}{2}\left(-0.3333-0.3333+b_1\left(\frac{4}{9}\right)-b_1\left(\frac{4}{9}\right)\right)*P_e$$
$$+(0.667)-\frac{l^2}{9}D\left(1-b_1\right),$$

$$E_{12} = \frac{l}{2}\left(0.3333-0.1667-b_2\left(\frac{4}{9}\right)-b_2\left(\frac{4}{9}\right)\right)*P_e$$
$$-(0.1667)-\frac{l^2}{18}D\left(1-2b_2\right),$$

$$E_{13} = \frac{l}{2}\left(0.1667+0.1667-b_3\left(\frac{4}{9}\right)+b_3\left(\frac{4}{9}\right)\right)*P_e$$
$$-(0.3333)-\frac{l^2}{36}D\left(1-4b_3\right),$$

$$E_{14} = \frac{l}{2}\left(-0.1667+0.3333+b_4\left(\frac{4}{9}\right)+b_4\left(\frac{4}{9}\right)\right)*P_e$$
$$-(0.1667)-\frac{l^2}{18}D\left(1-2b_4\right),$$

$$E_{21} = \frac{l}{2}\left(-0.3333-0.1667+b_1\left(\frac{4}{9}\right)-b_1\left(\frac{4}{9}\right)\right)*P_e$$
$$-(0.1667)-\frac{l^2}{18}D\left(1-2b_1\right),$$

$$E_{22} = \frac{l}{2}\left(0.3333-0.3333-b_2\left(\frac{4}{9}\right)-b_2\left(\frac{4}{9}\right)\right)*P_e$$
$$+(0.6667)+\frac{l^2}{9}D\left(1-b_2\right),$$

$$E_{23} = \frac{l}{2}\left(0.1667 + 0.3333 - b_3\left(\frac{4}{9}\right) + b_3\left(\frac{4}{9}\right)\right) * P_e$$
$$+ - (0.1667) - \frac{l^2}{18} D\,(1 - 2b_3)\,,$$
$$E_{24} = \frac{l}{2}\left(-0.1667 + 0.1667 + b_4\left(\frac{4}{9}\right) + b_4\left(\frac{4}{9}\right)\right) * P_e$$
$$- (0.3333) - \frac{l^2}{36} D\,(1 - 4b_4)\,,$$
$$E_{31} = \frac{l}{2}\left(-0.1667 - 0.1667 + b_1\left(\frac{4}{9}\right) - b_1\left(\frac{4}{9}\right)\right) * P_e$$
$$- (0.3333) - \frac{l^2}{36} D\,(1 - 4b_1)\,,$$
$$E_{32} = \frac{l}{2}\left(0.1667 - 0.3333 - b_2\left(\frac{4}{9}\right) - b_2\left(\frac{4}{9}\right)\right) * P_e$$
$$- (0.1667) - \frac{l^2}{18} D\,(1 - 2b_2)\,,$$
$$E_{33} = \frac{l}{2}\left(0.3333 + 0.3333 - b_3\left(\frac{4}{9}\right) + b_3\left(\frac{4}{9}\right)\right) * P_e$$
$$+ (0.6667) - \frac{l^2}{9} D\,(1 - b_3)\,,$$
$$E_{34} = \frac{l}{2}\left(-0.3333 + 0.1667 + b_4\left(\frac{4}{9}\right) + b_4\left(\frac{4}{9}\right)\right) * P_e$$
$$- (0.1667) - \frac{l^2}{18} D\,(1 - 2b_4)\,,$$
$$E_{41} = \frac{l}{2}\left(-0.1667 - 0.3333 + b_1\left(\frac{4}{9}\right) - b_1\left(\frac{4}{9}\right)\right) * P_e$$
$$- (0.1667) - \frac{l^2}{18} D\,(1 - 2b_1)\,,$$
$$E_{42} = \frac{l}{2}\left(0.1667 - 0.1667 - b_2\left(\frac{4}{9}\right) - b_2\left(\frac{4}{9}\right)\right) * P_e$$
$$- (0.3333) + \frac{l^2}{36} D\,(1 - 4b_2)\,,$$
$$E_{43} = \frac{l}{2}\left(0.3333 + 0.1667 - b_3\left(\frac{4}{9}\right) + b_3\left(\frac{4}{9}\right)\right) * P_e$$
$$- (0.1667) - \frac{l^2}{18} D\,(1 - 2b_3)\,,$$

$$E_{44} = \frac{l}{2}\left(-0.3333 + 0.3333 + b_4\left(\frac{4}{9}\right) + b_4\left(\frac{4}{9}\right)\right) * P_e$$
$$+ (0.6667) - \frac{l^2}{9} D\,(1 - b_4)\,.$$

Note that by insertion of zero for the bubble coefficients (E_{ij} $i, j = 1 \ldots, 4$) standard Galerkin solution of the problem is obtained. Any other value for the bubble coefficients introduces some degree of enrichment to the original shape functions. For example, using the values $l = 0.2, D_a = -10, P_e = 10$, and $b_1 = b_2 = b_3 = b_4 = 0.2325$, the terms of the ESM corresponding to enriched shape functions are found as

$$\mathrm{ESM} = \begin{bmatrix} 0.3412 & -0.0879 & 0.0079 & 0.3255 \\ -0.5478 & 0.8011 & 0.4522 & -0.1189 \\ -0.6589 & -0.4211 & 1.6744 & -0.0077 \\ -0.5478 & -0.5322 & 0.4522 & 1.2145 \end{bmatrix}.$$

After the assembly of the elemental stiffness matrices over the common nodes of the elements of the computational mesh, shown in Fig. 4.1 (25 square shape elements of 0.2×0.2 size) and imposition of the boundary conditions (Fig. 4.1), a determinate system of global equations is found.

Solution of the described system yields the nodal values of the field unknown in the interior nodes as

$$T = \begin{bmatrix} 0.0000, & 0.0000, & 0.0001, & 0.0018, & 0.0423, & 1.0000 \\ 0.0000, & 0.0000, & 0.0001, & 0.0018, & 0.0423, & 1.0000 \\ 0.0000, & 0.0000, & 0.0001, & 0.0018, & 0.0423, & 1.0000 \\ 0.0000, & 0.0000, & 0.0001, & 0.0018, & 0.0423, & 1.0000 \\ 0.0000, & 0.0000, & 0.0001, & 0.0018, & 0.0423, & 1.0000 \\ 0.0000, & 0.0000, & 0.0001, & 0.0018, & 0.0423, & 1.0000 \end{bmatrix}.$$

Inspection (via comparison with analogous one-dimensional results) provides a good indication for the reliability of the obtained solution.

4.2 Selection of Bubble Functions and Calculation of Bubble Coefficients

We now turn our attention to the important issue of the selection of appropriate bubble functions and evaluation of their corresponding bubble coefficients in two-dimensional problems. Consider the following two-dimensional transport equation. A solution for this equation over the square domain shown in Fig. 4.1 is sought. The field unknown, geometrical variables and

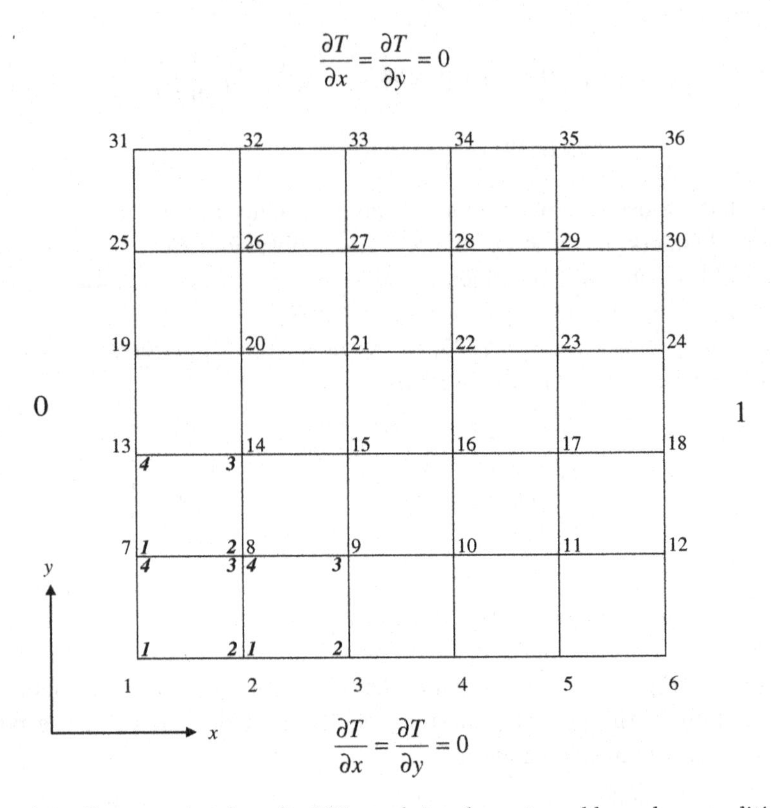

Fig. 4.1 Computational mesh of 25 equal size elements and boundary conditions.

the source term in the right-hand side of Eq. (4.7) are all made dimensionless using the relationships given in Chap. 3.

$$\left(\frac{\partial^2 T^*}{\partial x^{*2}} + \frac{\partial^2 T^*}{\partial y^{*2}}\right) - D_a T^* = f^*. \tag{4.7}$$

Equation (4.7) is solved using two different sets of boundary conditions. These are:

(a) BC1 — Physically representing a diffusion–reaction equation (DR) involving dissipation of the field variable within the problem domain

$$\begin{cases} T^* = 0 & \text{for } y^* = 0,\ 0 \le x^* \le 1 \text{ and } x^* = 0,\ 0 \le y^* < 1 \\ T^* = 1 & \text{for } x^* = 1,\ 0 \le y^* < 1 \text{ and } y^* = 1,\ 0 \le x^* \le 1 \end{cases} \tag{4.8}$$

and

(b) BC2 — Physically representing a DR equation involving production of the field variable within the problem domain

$$\begin{cases} T^* = 0 & \text{for } x^* = 0, \quad 0 \leq y^* \leq 1 \\ T^* = 1 & \text{for } x^* = 1, \quad 0 \leq y^* \leq 1 \\ \dfrac{\partial T^*}{\partial x^*} = \dfrac{\partial T^*}{\partial y^*} = 0 & \text{for } y^* = 0 \text{ and } y^* = 1 \text{ at } 0 < x^* < 1 \end{cases} \tag{4.9}$$

BC1 and BC2 can also be used for other transport equations.

4.3 RFB Functions Corresponding to the Two-Dimensional DR Equation

The first step is to solve the one-dimensional equivalent of Eq. (4.7) in order to obtain analytical one-dimensional shape functions (Chap. 3). Using a local coordinate system in a two-node linear element these are

$$\begin{cases} N_1 = \dfrac{\sinh\sqrt{D_a}(l - x)}{\sinh\sqrt{D_a}l} \\ N_2 = \dfrac{\sinh\sqrt{D_a}x}{\sinh\sqrt{D_a}l} \end{cases} . \tag{4.10}$$

As explained in Chap. 3, shape functions shown in set (4.10) can be expanded to construct polynomial functions such as

$$\begin{cases} \phi_1 = b_2(1-\xi^2) + b_3(1-\xi^2)(1-\xi) + b_4(1-\xi^2)^2 \\ \qquad + b_5(1-\xi^2)^2(1-\xi) + \cdots \\ \phi_2 = b_2(1-\xi^2) + b_3(1-\xi^2)(1+\xi) + b_4(1-\xi^2)^2 \\ \qquad + b_5(1-\xi^2)^2(1+\xi) + \cdots \end{cases} \tag{4.11}$$

or

$$\begin{cases} \phi_1 = b_2\varphi_{2,1}(\xi) + b_3\varphi_{3,1}(\xi) + b_4\varphi_{4,1}(\xi) + b_5\varphi_{5,1}(\xi) + \cdots \\ \phi_2 = b_2\varphi_{2,2}(\xi) + b_3\varphi_{3,2}(\xi) + b_4\varphi_{4,2}(\xi) + b_5\varphi_{5,2}(\xi) + \cdots \end{cases} \tag{4.12}$$

For bilinear elements, defined in Chap. 1, corresponding two-dimensional bubble functions are found by tensor products of the one-dimensional polynomials as

$$\begin{cases} \phi_1 = b_2\varphi_{2,1}(\xi)\varphi_{2,1}(\eta) + b_3\varphi_{3,1}(\xi)\varphi_{3,1}(\eta) + b_4\varphi_{4,1}(\xi)\varphi_{4,1}(\eta) \\ \qquad + b_5\varphi_{5,1}(\xi)\varphi_{5,1}(\eta) + \cdots \\ \phi_2 = b_2\varphi_{2,2}(\xi)\varphi_{2,1}(\eta) + b_3\varphi_{3,2}(\xi)\varphi_{3,1}(\eta) + b_4\varphi_{4,2}(\xi)\varphi_{4,1}(\eta) \\ \qquad + b_5\varphi_{5,2}(\xi)\varphi_{5,1}(\eta) + \cdots \\ \phi_3 = b_2\varphi_{2,2}(\xi)\varphi_{2,2}(\eta) + b_3\varphi_{3,2}(\xi)\varphi_{3,2}(\eta) + b_4\varphi_{4,2}(\xi)\varphi_{4,2}(\eta) \\ \qquad + b_5\varphi_{5,2}(\xi)\varphi_{5,2}(\eta) + \cdots \\ \phi_4 = b_2\varphi_{2,1}(\xi)\varphi_{2,2}(\eta) + b_3\varphi_{3,1}(\xi)\varphi_{3,2}(\eta) + b_4\varphi_{4,1}(\xi)\varphi_{4,2}(\eta) \\ \qquad + b_5\varphi_{5,1}(\xi)\varphi_{5,2}(\eta) + \cdots \end{cases} \tag{4.13}$$

Therefore, bubble-enriched shape functions which can be used in bilinear elements are derived by the combination of ordinary shape functions associated with these elements and the described two-dimensional tensor product functions. For example, shape functions corresponding to a two-dimensional third-order bubble-enriched bilinear element, which can be used to solve Eq. (4.7), are written as

$$\begin{cases} N_1 = \frac{1}{4}(1-\xi)(1-\eta) + b(1-\xi^2)(1-\eta^2)(3-\xi)(3-\eta) \\ N_2 = \frac{1}{4}(1+\xi)(1-\eta) + b(1-\xi^2)(1-\eta^2)(3+\xi)(3-\eta) \\ N_3 = \frac{1}{4}(1+\xi)(1+\eta) + b(1-\xi^2)(1-\eta^2)(3+\xi)(3+\eta) \\ N_4 = \frac{1}{4}(1-\xi)(1+\eta) + b(1-\xi^2)(1-\eta^2)(3-\xi)(3+\eta) \end{cases}, \tag{4.14}$$

where $b = \frac{-1}{8}\left(1 + \frac{6}{D_a l^2}\right)^{-1}$, in which l is a characteristic element dimension (length).

After the derivation of the bubble-enriched shape functions, weighted residual statement of the governing Eq. (4.7) can be constructed in the normal manner as

$$\int_{\Omega_e} \left[W_i \left(\frac{\partial^2 \sum_{j=1}^{n} N_j T_j^*}{\partial x^{*2}} + \frac{\partial^2 \sum_{j=1}^{n} N_j T_j^*}{\partial y^{*2}} \right) \right. \\ \left. - D_a\, W_i \sum_{j=1}^{n} N_j T_j^* - W_i f^\circ \right] dx dy = 0. \tag{4.15}$$

To develop a solution procedure based on Eq. (4.15), the weights (W_i) are kept identical to the ordinary Lagrangian shape functions whilst the shape functions (N_i) are enriched by the bubble functions (e.g. functions given by equation set of (4.14)).

4.3.1 *Derivation of polynomial bubble function coefficients for the DR equation using the STC method*

Samples of polynomial functions which can be selected to enrich ordinary shape functions of linear elements in the STC method are introduced in Chap. 3. As mentioned earlier two-dimensional "bilinear" equivalent of enriched one-dimensional shape functions can be constructed using tensor products of one-dimensional functions as

$$N_i(\xi, \eta) = \psi_i(\xi, \eta) + b\phi(\xi)\phi(\eta), \tag{4.16}$$

where $\psi_i(\xi, \eta)$ is bilinear Lagrangian shape function, b is a bubble coefficient (calculated via STC) and $\phi(\xi)$ and $\phi(\eta)$ are bubble functions in terms of one-dimensional space variables ξ and η, respectively.

4.3.2 *Elimination of the boundary integrals*

The way in which the boundary integrals, which appear after the application of Green's theorem to the weighted residual statement (e.g. Eq. (4.15)), are eliminated in multidimensional cases is different from the one used for one-dimensional problems. In general, in discretizations involving bubble functions the inter-element boundary integrals are not automatically eliminated during the assembly of elemental stiffness equations into a global system. This problem does not occur in a one-dimensional case as the boundary integrals are reduced to simple nodal flux terms. However, if we consider the variational formulation of the DR equation, for example, after the application of Green's theorem we have

$$(\nabla T, \nabla v) + (D_a T_h, v) = (f, v). \tag{4.17}$$

Substitution from $T = T_c + T_b$ (Eq. (3.23)) gives

$$(\nabla T_c, \nabla v) + (\nabla T_b, \nabla v) + (D_a T, v) = (f, v). \tag{4.18}$$

If v is selected to be a linear test function (weight function) then according to Green's theorem (Franca and Farhat, 1995), we have

$$(\nabla v, \nabla \phi)_{\Omega_e} = -(\Delta v, \phi)_{\Omega_e} + (\nabla v, \phi)_{\Gamma_e} = 0, \tag{4.19}$$

where ϕ is a bubble function. Therefore, Eq. (4.18) is reduced to

$$(\nabla T_c, \nabla v) + (D_a T_h, v) = (f, v). \tag{4.20}$$

As can be seen, the bubble function does not affect the Laplacian term in the DR equation and therefore no boundary integral due to the bubble function exists.

4.3.3 *Solution of a benchmark two-dimensional DR problem*

In this section the solution of a benchmark problem using a rectangular domain discretized into an equal density mesh of 100 elements is shown in Fig. 4.2 (here this mesh scheme is designated as BLNR1). Two sets of numerical solutions, one based on the standard Galerkin method based on ordinary bilinear elements and the other using bubble-enriched elements are obtained. To validate the numerical solutions, they are compared with the following analytical solution of the dimensionless DR equation. The analytical solution is obtained in conjunction with the set of boundary

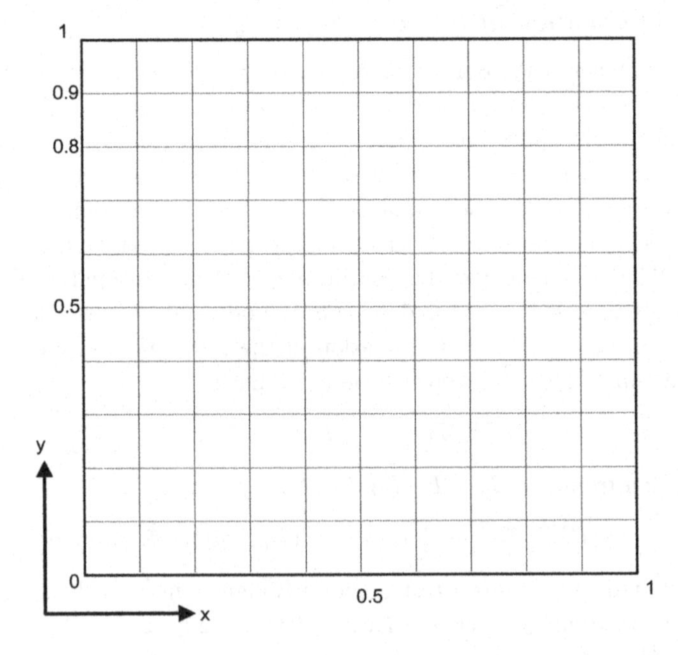

Fig. 4.2 10 × 10 uniform density mesh of square shape elements BLNR1.

conditions given by Eq. (4.8) using the method of the separation of variables (Parvazinia and Nassehi, 2006).

$$T^*(x^*, y^*) = \frac{2}{\pi}\sum_{n=1}^{\infty}\beta_n \frac{1-(-1)^n}{n\ \sinh\sqrt{\gamma}}\{\sinh\sqrt{\gamma}y^* \sin(n\pi x^*) + \sinh\sqrt{\gamma}x^* \sin(n\pi y^*)\}, \tag{4.21}$$

where $\gamma = \sqrt{n^2\pi^2 + D_a}$.

Using the boundary conditions represented by Eq. (4.9), a one-dimensional solution is found as

$$T^*(x^*) = \frac{\sin(\sqrt{D_a}x^*)}{\sin(\sqrt{D_a})}. \tag{4.22}$$

Using mesh BLNR1 and boundary conditions BC1 (Eq. (4.8)), the elemental and global stiffness matrix corresponding to the standard Galerkin method based on ordinary bilinear elements are calculated as

$$\mathrm{ESM} = \begin{bmatrix} 1.5 & 0.25 & -0.125 & 0.25 \\ 0.25 & 1.5 & 0.25 & -0.125 \\ -0.125 & 0.25 & 1.5 & 0.25 \\ 0.25 & -0.125 & 0.25 & 1.5 \end{bmatrix}.$$

The global 81 × 81 matrix and its corresponding load vector are

$$\begin{bmatrix}
6.0001 & 0.4999 & 0 & 0 & 0 & 0 & 0 & 0 & \cdots & \cdots 0 \\
0.4999 & 6.0001 & 0.4999 & 0 & 0 & 0 & 0 & 0 & \cdots & \cdots 0 \\
0 & 0.4999 & 6.0001 & 0.4999 & 0 & 0 & 0 & 0 & \cdots & \cdots 0 \\
0 & 0 & 0.4999 & 6.0001 & 0.4999 & 0 & 0 & 0 & \cdots & \cdots 0 \\
0 & 0 & 0 & 0.4999 & 6.0001 & 0.4999 & 0 & 0 & \cdots & \cdots 0 \\
0 & 0 & 0 & 0 & 0.4999 & 6.0001 & 0.4999 & 0 & \cdots & \cdots 0 \\
0 & 0 & 0 & 0 & 0 & 0.4999 & 6.0001 & 0.4999 & 0 & \cdots 0 \\
0 & 0 & 0 & 0 & 0 & 0 & 0.4999 & 6.0001 & 0.4999 & 0 \cdots 0 \\
\cdots & \cdots & \cdots & \cdots & \cdots & \cdots & \cdots & \cdots & \cdots & \\
\cdots & \cdots & \cdots & \cdots & \cdots & \cdots & \cdots & \cdots & \cdots & \cdots \\
0 & \cdots & 0 & 0 & 0 & 0 & 0 & 0.4999 & 6.0001 & 0.4999 \\
0 & \cdots & 0 & 0 & 0 & 0 & 0 & 0 & 0.4999 & 6.0001
\end{bmatrix},$$

$$
\begin{bmatrix}
0 \\ 0 \\ 0 \\ 0 \\ 0 \\ 0 \\ 0 \\ 0 \\ -0.25 \\ \vdots \\ -0.375 \\ -0.25 \\ -0.25 \\ -0.25 \\ -0.25 \\ -0.25 \\ -0.25 \\ -0.25 \\ -0.625
\end{bmatrix}.
$$

The solution obtained in this case is represented by the following matrix

$$
\begin{bmatrix}
0.0000, & 0.0000, & 0.0000, & 0.0000, & 0.0000, & 0.0000, & -0.0002, & 0.0024, & -0.0389 \\
0.0000, & 0.0000, & 0.0000, & 0.0000, & 0.0000, & 0.0000, & 0.0000, & 0.0011, & -0.0355 \\
0.0000, & 0.0000, & 0.0000, & 0.0000, & 0.0000, & 0.0000, & -0.0001, & 0.0013, & -0.0358 \\
0.0000, & 0.0000, & 0.0000, & 0.0000, & 0.0000, & 0.0000, & 0.0000, & 0.0013, & -0.0358 \\
0.0000, & 0.0000, & 0.0000, & 0.0000, & 0.0000, & 0.0000, & 0.0000, & 0.0013, & -0.0358 \\
0.0000, & 0.0000, & 0.0000, & 0.0000, & 0.0000, & 0.0000, & 0.0000, & 0.0012, & -0.0357 \\
-0.0004, & 0.0001, & -0.0001, & 0.0000, & 0.0000, & 0.0000, & -0.0002, & 0.0014, & -0.0362 \\
0.0043, & 0.0003, & 0.0014, & 0.0013, & 0.0013, & 0.0012, & 0.0014, & 0.0014, & -0.0311 \\
-0.0600, & -0.0336, & -0.0360, & -0.0357, & -0.0358, & -0.0357, & -0.0362, & -0.0311, & -0.0989
\end{bmatrix}.
$$

The first row represents the nodal values of the unknown at $y^* = 0.1$, the second row at $y^* = 0.2$, and so on.

The presented results correspond to $D_a = 750$ for a dissipative reaction case subject to boundary conditions (4.8). Sample results obtained using bubble-enriched and ordinary bilinear elements are compared with the analytical solution in Fig. 4.3. As Figs. 4.3(a) and 4.3(b) show, at $D_a = 750$ a fifth-order RFB function is capable of producing stable–accurate solution which is comparable with one-dimensional solution if the same order bubble coefficients are used. At cross sections indicated by $y^* = 0.5$ and $y^* = 0.9$ for the last row of elements at x^*=0.9, results generated by the fifth-order

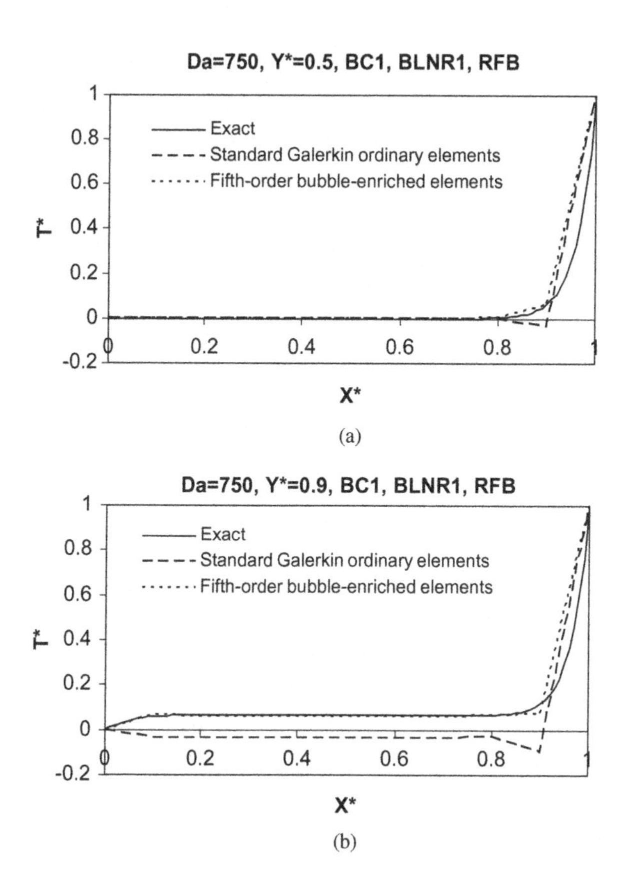

Fig. 4.3 The results for $D_a = 750$, mesh schemes BLNR1 and boundary conditions BC1 with RFB method, dissipation case.

RFB function show only a slight inaccuracy whilst the solution obtained using ordinary elements is quite inaccurate and unstable.

By using an alternative 5×5 uniform density mesh (here designated as mesh scheme BLNR2) the ordinary elements-based Galerkin solution shows strong instability whilst fifth-order RFB function remains mainly stable, even in such a coarse mesh (Fig. 4.4). However, by increasing the order of the bubble function to seven, totally stable and accurate results are found using mesh scheme BLNR2.

The solutions presented in Figs. 4.3 and 4.4 are found via the RFB function method and therefore, depend on the availability of analytical results. It should, however, be noted that two-dimensional bubble functions are found as the tensor products of one-dimensional functions. Therefore,

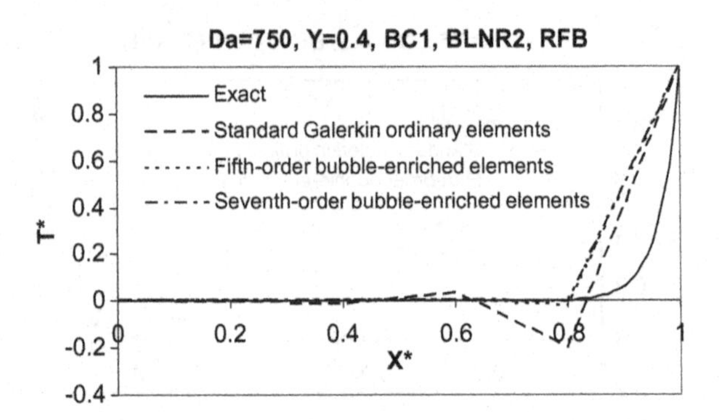

Fig. 4.4 Comparison of the analytical solution with results obtained using standard Galerkin and RFB methods ($D_a = 750$, mesh scheme BLNR2, and boundary conditions BC1, dissipation case).

the analytical solutions obtained for analogous one-dimensional problems are used to construct two-dimensional bubble functions. This is true even in cases, such as the present problem, in which a two-dimensional analytical solution can be found if specific types of boundary conditions are prescribed.

The benchmark problem presented in this example can also be solved using the STC bubble function method whose concept and procedure are explained previously in this book. As shown in Fig. 4.5, STC-based bubble functions of sixth- and eighth-order generate stable and accurate results on the 10×10 mesh (BLNR1) used here.

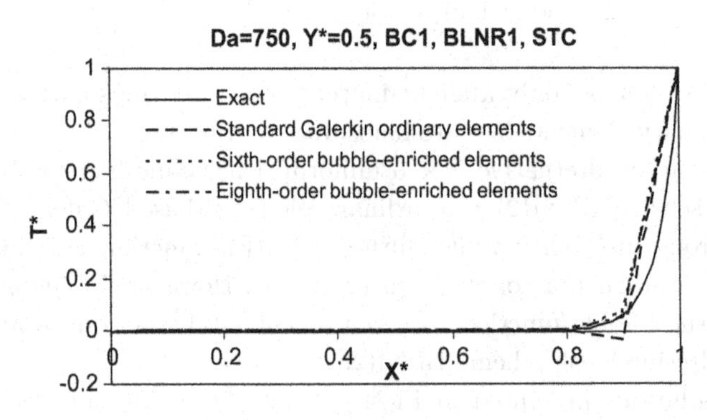

Fig. 4.5 Comparison of the analytical solution with results obtained using standard Galerkin and STC methods ($D_a = 750$, mesh scheme BLNR1, BC1, dissipation case).

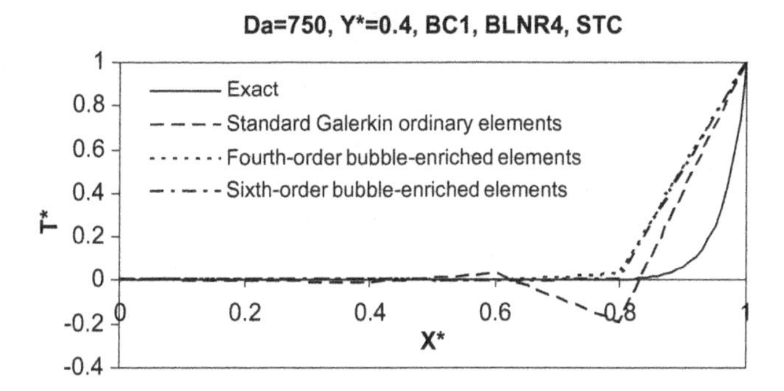

Fig. 4.6 Comparison of the analytical solution with results obtained using standard Galerkin and STC methods ($D_a = 750$, mesh scheme BLNR2, BC1, dissipation case).

Fourth- and sixth order STC-based bubble functions (Fig. 4.6) are also capable of generating stable and accurate results on the 5×5 element mesh (BLNR2).

The results obtained for the above described two-dimensional DR equation show that, in general, bubble-enriched elements provide robust approximations in the solution of dissipative reaction problems. Even solutions obtained using very coarse meshes show a good degree of accuracy within the domain and near its boundaries. This provides a clear indication that the described methods for the construction of two-dimensional bubble-enriched shape functions and elimination of boundary integrals are valid and can be used to develop practical schemes for the solution of multiscale transport problems of this type.

We now consider productive reaction problems. In this case the two-dimensional DR equation is solved in conjunction with boundary conditions BC2 shown in Eq. (4.9). A $D_a = -750$ is inserted into the governing equation and initially we use the 10×10 mesh (BLNR1). In contrast to the dissipative case, where the multiscale behavior is limited to a thin near-wall-layer, in productive reaction problems the multiscale behavior is repeated at inter-layers over the entire domain. In this case the application of the standard Galerkin method in conjunction with ordinary bilinear elements fails if a coarse mesh is used and in general, generates unstable solutions. However, after a four-fold mesh refinement and using a mesh of 40×40 (BLNR3) the standard Galerkin scheme generates a stable result. In Fig. 4.7 this solution is compared with the exact analytical solution and the result obtained

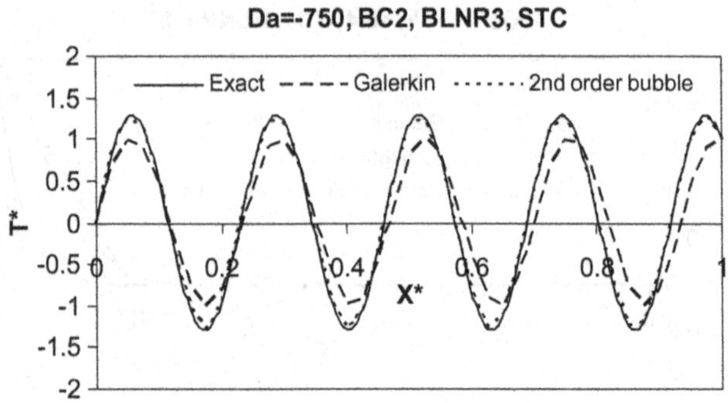

Fig. 4.7 Comparison of the analytical, standard Galerkin, and second-order bubble function results ($D_a = -750$, mesh scheme BLNR3, BC2, Production case).

using the finite element scheme based on second-order bubble function-enriched method. As shown in this figure, the bubble function-based scheme generates a significantly more accurate result than the scheme which uses ordinary bilinear elements. More detailed discussion of these points can be found in Parvazinia and Nassehi (2007).

4.4 Solution of Two-Dimensional Convection–Diffusion (CD) Equation

In this section bubble function-enriched finite element solution of CD equation is described. Historically, the solution of this class of equations has presented many challenges for numerical analysts and research in this area has led to the development of powerful techniques such as previously presented Streamline Upwind Petrov–Galerkin (SUPG) technique. Application of this scheme may, however, require the use of complicated algorithms based on C^1 continuous finite elements (Nassehi, 2002). Finite element schemes based on bubble function-enriched elements can resolve such problems and relatively simple schemes can be used to obtain stable and accurate solutions for this class of governing equations which arise in many types of practical transport phenomena.

4.4.1 *Governing CD equation and boundary conditions*

Using vector notations steady state homogeneous CD equation can be represented as

$$\mathrm{v}.\nabla T - k\nabla.\nabla T = f, \tag{4.23}$$

where T is a field unknown, $\mathbf{v}$ is the velocity vector, k is diffusivity and f is a source term. ∇ denotes the spatial gradient operator. Using the following relationships a non-dimensional form of Eq. (4.23) is derived.

$$\begin{cases} T^* = \dfrac{T}{T_0} \\ x^* = \dfrac{x}{h},\ y^* = \dfrac{y}{h} \end{cases}, \tag{4.24}$$

where T_0 is a reference value for the field variable (e.g. temperature), and h is a characteristic length (e.g width of the domain), therefore,

$$p_e \nabla T^* - \nabla . \nabla T^* = f^*, \tag{4.25}$$

in which P_e is the Peclet number defined by $P_e = \frac{vh}{k}$ and $f^* = f\frac{h^2}{kT_0}$.

Based on Eq. (4.25), in a planar (x, y) coordinate system two-dimensional CD equation is expressed in a non-dimensional form as

$$\left(p_{ex} \frac{\partial T^*}{\partial x^*} + p_{ey} \frac{\partial T^*}{\partial y^*} \right) - \left(\frac{\partial^2 T^*}{\partial x^{*2}} + \frac{\partial^2 T^*}{\partial y^{*2}} \right) = f^*, \tag{4.26}$$

where p_{ex} and p_{ey} are defined using the x and y components of the velocity vector. However, Peclet number in both directions is assumed to have the same numerical value. Equation (4.26) is assumed to be subject to the set of boundary conditions shown by Eq. (4.9) (i.e. BC2) over the domain shown in Fig. 4.1.

4.4.2 *RFB functions corresponding to the two-dimensional CD equation*

Similar to the case of DR equation the first step here is to solve the one-dimensional equivalent of Eq. (4.26) in order to obtain analytical one-dimensional shape functions (Chap. 3). Using a local coordinate system defined over a two-node linear element these are

$$\begin{cases} N_1 = \dfrac{\exp(p_e x) - \exp(p_e l)}{1 - \exp(p_e l)} \\ N_2 = \dfrac{1 - \exp(p_e x)}{1 - \exp(p_e l)} \end{cases}. \tag{4.27}$$

4.4.3 *Two-dimensional bubble functions for bilinear elements*

Equation set of (4.27) provides the basis for the derivation of various order two-dimensional bubble-enriched shape functions as tensor products of one-dimensional functions for both RFB and STC schemes. Details of this derivation are explained in Chap. 3 and here samples of bubble functions used to solve the two-dimensional CD equation are presented.

Second-order bubble functions:

$$\begin{cases} N_1 = \frac{1}{4}(1-\xi)(1-\eta) + b(1-\xi^2)(1-\eta^2) \\ N_2 = \frac{1}{4}(1+\xi)(1-\eta) + b(1-\xi^2)(1-\eta^2) \\ N_3 = \frac{1}{4}(1+\xi)(1+\eta) + b(1-\xi^2)(1-\eta^2) \\ N_4 = \frac{1}{4}(1-\xi)(1+\eta) + b(1-\xi^2)(1-\eta^2) \end{cases}. \tag{4.28}$$

Fourth-order bubble functions:

$$\begin{cases} N_1 = \frac{1}{4}(1-\xi)(1-\eta) + b_1(1-\xi^2)(1-\eta^2) \\ \quad + b_2(1+\xi)(1+\eta)(1-\xi^2)(1-\eta^2) + b_3(1-\xi^2)^2(1-\eta^2)^2 \\ N_2 = \frac{1}{4}(1+\xi)(1-\eta) + b_1(1-\xi^2)(1-\eta^2) \\ \quad + b_2(1+\xi)(1+\eta)(1-\xi^2)(1-\eta^2) - b_3(1-\xi^2)^2(1-\eta^2)^2 \\ N_3 = \frac{1}{4}(1+\xi)(1+\eta) + b_1(1-\xi^2)(1-\eta^2) \\ \quad + b_2(1+\xi)(1+\eta)(1-\xi^2)(1-\eta^2) + b_3(1-\xi^2)^2(1-\eta^2)^2 \\ N_4 = \frac{1}{4}(1-\xi)(1+\eta) + b_1(1-\xi^2)(1-\eta^2) \\ \quad + b_2(1+\xi)(1+\eta)(1-\xi^2)(1-\eta^2) - b_3(1-\xi^2)^2(1-\eta^2)^2 \end{cases}. \tag{4.29}$$

Bubble coefficients appearing in Eqs. (4.28) and (4.29) are found using procedures similar to those explained for one-dimensional functions.

4.4.4 *Elimination of the boundary integrals*

The procedure that can be used to eliminate boundary integrals appearing after the application of Green's theorem to the Laplacian term in the CD equation is similar to the one explained in Sec. 4.3.2 for the DR equation. Considering the variational form of CD equation, given as

$$(P_e \nabla T, v) + (\nabla T_1, \nabla v) = (f, v), \tag{4.30}$$

it can be seen that the bubble function does not affect the Laplacian term in the CD equation and therefore no boundary integral due to the bubble function appears after the application of Green's theorem.

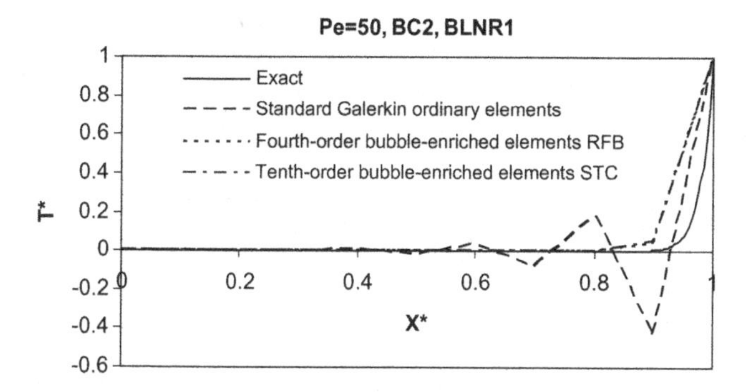

Fig. 4.8 Comparison of the analytical, standard Galerkin, and fourth- and tenth-order bubble functions schemes ($P_e = 50$, mesh scheme BLNR1, BC2).

4.4.5 *Solution of a benchmark two-dimensional CD problem*

Using a mesh similar to Fig. 4.2, the CD equation is solved subject to boundary conditions (4.9). To validate the numerical solutions, the analytical solution of the dimensionless CD equation has been used. In Fig. 4.8 solutions obtained using analytical, standard Galerkin based on ordinary elements and fourth- and tenth-order bubble function-enriched RFB and STC methods are compared. The results shown in Fig. 4.8 correspond to nodes located along the diagonal line starting from position $x = y = 0$ in the square domain used in this problem. As can be seen from Fig. 4.8, in contrast to the stable results generated by the bubble function schemes the standard Galerkin result is oscillatory and unstable.

After a three-fold increase of the Peclet number, the standard Galerkin scheme generates oscillatory results all over the domain whilst the bubble-enriched scheme produces stable and accurate results over the nodes of the computational mesh. In Fig. 4.9 the results corresponding to the diagonal line for three cases of analytical, standard Galerkin based on ordinary elements and fourth-order bubble function-enriched RFB methods are shown.

4.5 Solution of Convection–Diffusion–Reaction (CDR) Problems

CDR equation includes all three transport mechanisms for a property over a physical domain. In general, analytical solution of this equation is not possible and it can only be solved using appropriate numerical schemes. However, similar to many existing techniques, the standard Galerkin method based

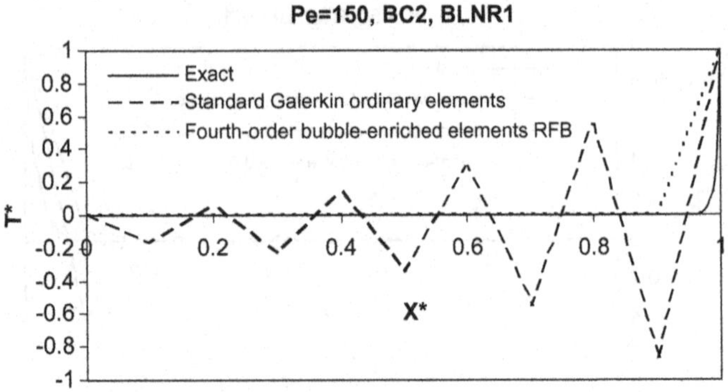

Pe=150, BC2, BLNR1

Fig. 4.9 Comparison of the analytical, standard Galerkin, and fourth-order bubble functions schemes ($P_e = 150$, BLNR1, BC2).

on ordinary types of finite elements fails to generate stable results under convection or reaction dominated situations for this equation. In this section, we develop bubble function-enriched finite element schemes for the CDR equation and evaluate their performance under various conditions.

4.5.1 *Governing CDR equation and boundary conditions*

Using vector notations steady state CDR equation in domain $\Omega \subset R^d$ can be written as

$$\mathbf{v}.\nabla T - k\nabla.\nabla T - sT = f, \tag{4.31}$$

where T is the field unknown, $\mathbf{v}$ is the velocity vector, k is the diffusion coefficient, $s > 0$ indicates productive reaction and $s < 0$ stands for dissipative reaction cases, and f is a given source term. Using the following dimensionless forms

$$\begin{cases} T^* = \dfrac{T}{T_0} \\ \bar{x}^* = \dfrac{\bar{x}}{h} \end{cases}, \tag{4.32}$$

where superscript * represents a dimensionless variable, T_0 is a reference value for the field variable, h is a characteristic length (e.g. width of the domain) and $\bar{x}$ represents position vector in the selected coordinate system. After substitution from Eq. (4.32) the general CDR equation is written in a dimensionless form as

$$\nabla T^* - \frac{1}{P_e}\nabla.\nabla T^* - D_a T^* = f^*, \tag{4.33}$$

in which f^* is the dimensionless source term, P_e is the Peclet number and D_a is the Damköhler number, respectively, defined as

$$\begin{cases} P_e = \dfrac{vh}{k} \\ D_a = \dfrac{sh}{v} \\ f^* = \dfrac{h}{T_0 v} f \end{cases} . \tag{4.34}$$

It has to be noted that Eq. (4.31) should be made non-dimensional in a way that the resulting equation can directly be written as $P_e\nabla T^* - \nabla.\nabla T^* - D_a T^* = f^*$, where $D_a = \frac{sh^2}{k}$ and $f^* = f\frac{h^2}{kT_0}$. This means that the Damköhler number which generally represents the reaction rate to the transport rate (or dynamic rate) has different definitions in the DR and CDR equations used in this book. This is in accordance with the generally accepted description of this dimensionless number (Brodkey and Hershey, 1988). It is also assumed that P_e and D_a are the same along all coordinate directions. In a two-dimensional system (x^*, y^*) Eq. (4.33) can hence be written as

$$\left(\frac{\partial T^*}{\partial x^*} + \frac{\partial T^*}{\partial y^*}\right) - \frac{1}{P_e}\left(\frac{\partial^2 T^*}{\partial x^{*2}} + \frac{\partial^2 T^*}{\partial y^{*2}}\right) - D_a T^* = f^*. \tag{4.35}$$

Equation (4.35) can now be solved using the following sets of boundary conditions prescribed along the external boundaries of a rectangular domain.

(a) Exponential regime (BC1):

$$\begin{cases} T^* = 0 & \text{for } y^* = 0,\ 0 \le x^* \le 1 \text{ and } x^* = 0,\ 0 \le y^* < 1 \\ T^* = 1 & \text{for } x^* = 1,\ 0 \le y^* < 1 \text{ and } y^* = 1,\ 0 \le x^* \le 1 \end{cases} \tag{4.36}$$

and

(b) Propagation regime (BC2):

$$\begin{cases} T^* = 0 & \text{for } x^* = 0,\quad 0 \le y^* \le 1 \\ T^* = 1 & \text{for } x^* = 1,\quad 0 \le y^* \le 1 \\ \dfrac{\partial T^*}{\partial x^*} = \dfrac{\partial T^*}{\partial y^*} = 0 & \text{for } y^* = 0 \text{ and } y^* = 1 \text{ at } 0 < x^* < 1. \end{cases} \tag{4.37}$$

4.5.2 *Derivation of RFB functions for the CDR equation*

Similar to the DR and CD equations, in this case again analogous one-dimensional CDR equation is solved to obtain the following shape functions:

$$\begin{cases} N_1 = \dfrac{\begin{array}{c}\exp((0.5P_e - \alpha)l + (0.5P_e + \alpha)x) \\ - \exp((0.5P_e + \alpha)l + (0.5P_e - \alpha)x)\end{array}}{\exp(0.5P_e - \alpha)l - \exp(0.5P_e + \alpha)l}, \\ N_2 = \dfrac{\exp(0.5P_e - \alpha)x - \exp(0.5P_e + \alpha)x}{\exp(0.5P_e - \alpha)l - \exp(0.5P_e + \alpha)l} \end{cases} \tag{4.38}$$

where

$$\alpha = 0.5\sqrt{P_e^2 - 4P_e D_a}. \tag{4.39}$$

Two-dimensional shape functions are derived using tensor products of one-dimensional shape functions as shown for the DR and CD equations. The bubble function coefficients remain the same as those obtained for the one-dimensional problem.

Elimination of boundary integrals in the solution of the CDR problem is similar to the procedure explained in the case of the DR equation. The weighted residual statement of the problem is given as

$$(\nabla T_h, v) + \left(\frac{1}{P_e}\nabla T_c, \nabla v\right) - (D_a T_h, v) = (f, v). \tag{4.40}$$

Equation (4.40) shows that the use of a bubble function does not affect the Laplacian term in the CDR equation and therefore no boundary integral term is generated because of its use in the solution of the CDR equation.

4.5.3 *Solution of a benchmark two-dimensional CDR equation*

Using a 10 $\times$ 10 mesh similar to the one shown in Fig. 4.2, two sets of results for the CDR equation are obtained. The first solution corresponds to the boundary conditions given in Eq. (4.36) and the second set of results is obtained using boundary conditions given in Eq. (4.37). To evaluate the level of accuracy of the numerical results they are compared with the analytical solution of the dimensionless CDR equation. The analytical solution for the exponential regime is based on the method of separation of variables

(Parvazinia and Nassehi, 2010) and is represented as

$$T^*(x^*, y^*) = \exp\left(p_e\left(\frac{x^*+y^*}{2}\right)\right)\sum_{n=1}^{\infty}\beta_n\frac{1}{1-e^{-2\alpha}} \times \left\{\sin(n\pi x^*)\left(e^{-\frac{\alpha}{2}(1-y^*)} - e^{-\frac{\alpha}{2}(1+y^*)}\right) + \sin(n\pi y^*)\left(e^{-\frac{\alpha}{2}(1-y^*)} - e^{\frac{-\alpha}{2}(1+y^*)}\right)\right\}, \quad (4.41)$$

where $\alpha = \sqrt{P_e^2 + 4n^2\pi^2 - 4D_aP_e}$ and

$$\beta_n = \frac{8n\pi\left\{1-(-1)^n\exp\left(\frac{-p_e}{2}\right)\right\}}{p_e^2 + 4n^2\pi^2}.$$

For the propagation case defined by the boundary conditions given in Eq. (4.37), the following analogous one-dimensional solution can be obtained

$$T^*(x^*) = \frac{\exp(0.5P_ex^*)\sin(\sqrt{|P_e^2 - 4P_eD_a|}x^*)}{\exp(0.5P_e)\sin(\sqrt{|P_e^2 - 4P_eD_a|})}. \quad (4.42)$$

Inserting values of P_e=10 and $D_a = -10$, the standard Galerkin scheme based on ordinary bilinear elements is carried out and its results are compared with the solution obtained using bubble function-enriched schemes. As shown in Fig. 4.10, for this set of values the standard Galerkin scheme generates results which are very similar to the set of results obtained using a

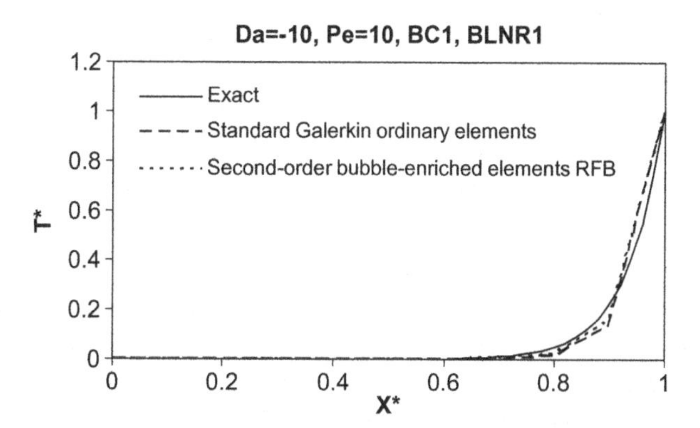

Fig. 4.10 Comparison of the analytical, standard Galerkin, and second-order bubble function schemes ($P_e = 10$ and $D_a = -10$ at $y^* = 0.5$, mesh scheme BLNR1, exponential regime).

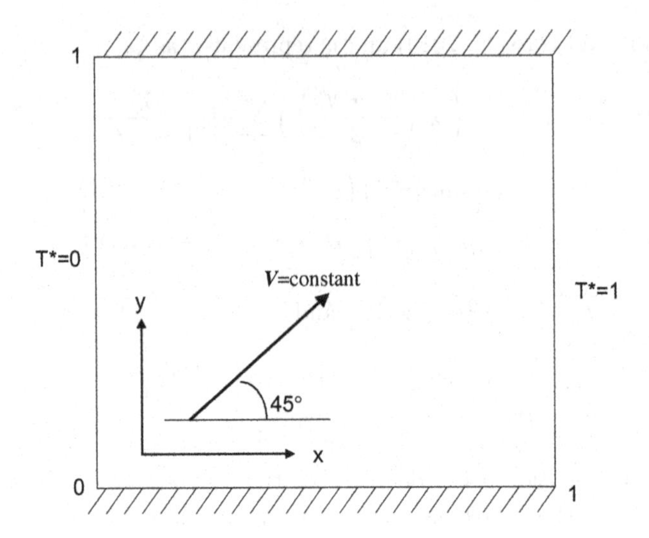

Fig. 4.11 Boundary conditions used to solve CDR (exponential regime).

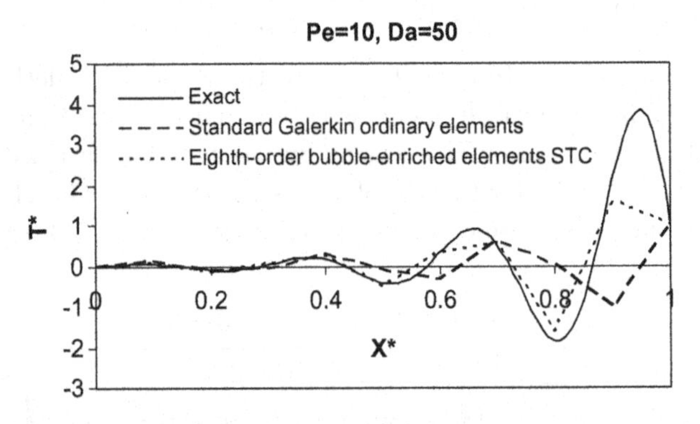

Fig. 4.12 Comparison of the analytical, standard Galerkin, and eighth-order bubble function, STC schemes (P_e = 10 and D_a = 50, mesh scheme BLNR1, propagation regime).

second-order bubble function-enriched RFB scheme. Both sets of numerical results are also in close agreement with the analytical solution (Fig. 4.11).

Using values of $P_e = 10$ and $D_a = 50$, performance of the standard Galerkin and bubble-enriched schemes in a propagation regime are compared. In this case, as shown in Fig. 4.12, standard Galerkin scheme fails to yield accurate results. In contrast, eighth-order bubble function-enriched STC scheme generates acceptable nodal solutions.

4.6 Solution of Transport Equations Using Bubble Function-Enriched Triangular Elements

Triangular elements offer greater geometrical flexibility than quadrilateral elements. Construction of bubble-enriched shape functions for quadrilateral elements is straightforward because they can be formed via tensor products of one-dimensional functions. However, a class of bubble functions for triangular elements can also be readily constructed. In this section, construction of a scheme based on triangular bubble-enriched elements for the solution of DR equation is outlined.

Consider the following set of shape functions defined using local coordinate system (ξ, η) corresponding to a "linear" triangular element (see Chap. 2, Sec. 2.2.2):

$$\begin{cases} N_1 = (1 - \xi - \eta) \\ N_2 = \xi \\ N_3 = \eta \end{cases} . \tag{4.43}$$

A bubble function based on these functions can be constructed as

$$\phi_b = N_1 N_2 N_3 = \xi\eta(1 - \xi - \eta). \tag{4.44}$$

Higher-order bubble functions for a triangular element can also be written as (Parvazinia and Nassehi, 2007)

$$\begin{aligned} \phi_b &= \xi\eta(1 - \xi - \eta) + (\xi\eta(1 - \xi - \eta))^2 + (\xi\eta(1 - \xi - \eta))^3 + \cdots \\ &= \sum_{q=1}^{n} (\xi\eta(1 - \xi - \eta))^q. \end{aligned} \tag{4.45}$$

After application of the STC method, the bubble coefficient corresponding to these functions can be derived as follows.

Second-order bubble function:

$$\begin{cases} \phi_b = \xi\eta(1 - \xi - \eta) \\ b = \dfrac{-1}{\dfrac{2}{D_a}\left(\dfrac{1}{l_x^2} + \dfrac{1}{l_y^2}\right) + 0.071429} \end{cases} . \tag{4.46}$$

Fourth-order bubble function:

$$\begin{cases} \phi_b = \xi\eta(1 - \xi - \eta) + (\xi\eta(1 - \xi - \eta))^2 \\ b = \dfrac{-1}{\dfrac{1}{D_a}\left(\dfrac{1}{l_x^2} + \dfrac{1}{l_y^2}\right) + 0.124368} \end{cases} , \tag{4.47}$$

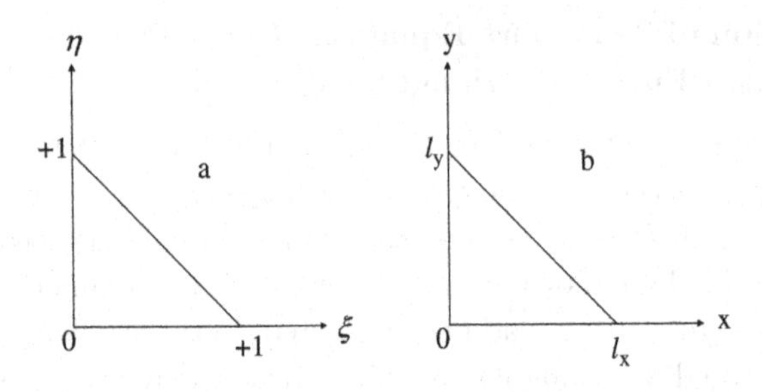

Fig. 4.13 Triangular master element represented using local normalized (ξ, η) and local (x, y) coordinate systems.

where l_x and l_y are characteristic element length in the x and y direction of Cartesian coordinate system.

To construct bubble function-enriched shape functions for the "linear" triangular element, the described bubble functions are added to the ordinary Lagrangian shape functions derived in Chap. 2. Therefore, the second-order bubble-enriched shape functions corresponding to enriched triangular element are

$$\begin{cases} N_1 = (1 - \xi - \eta) + b\xi\eta(1 - \xi - \eta) \\ N_2 = \xi + b\xi\eta(1 - \xi - \eta) \\ N_3 = \eta + b\xi\eta(1 - \xi - \eta) \end{cases}, \tag{4.48}$$

where b is the bubble coefficient defined in Eq. (4.46). The master element corresponding to these shape functions is shown in Figs. 4.13(a) and 4.13(b).

We now consider the solution of Eq. (4.35) over the mesh shown in Fig. 4.14, subject to the boundary conditions shown in this figure.

Following the normal procedure of weighted residual finite element method, a variational statements is constructed via weighting of the residual obtained by the substitution of the field unknown in terms of shape functions in the differential equation. After the application of Green's theorem to second-order derivatives, this statement can be used to obtain the required elemental stiffness equation. Therefore, for the DR equation we have

$$\int_0^{l_x} \int_0^{l_y - \frac{l_y}{l_x}x} ((T_x w_x + T_y w_y) + D_a w T) dy dx = 0. \tag{4.49}$$

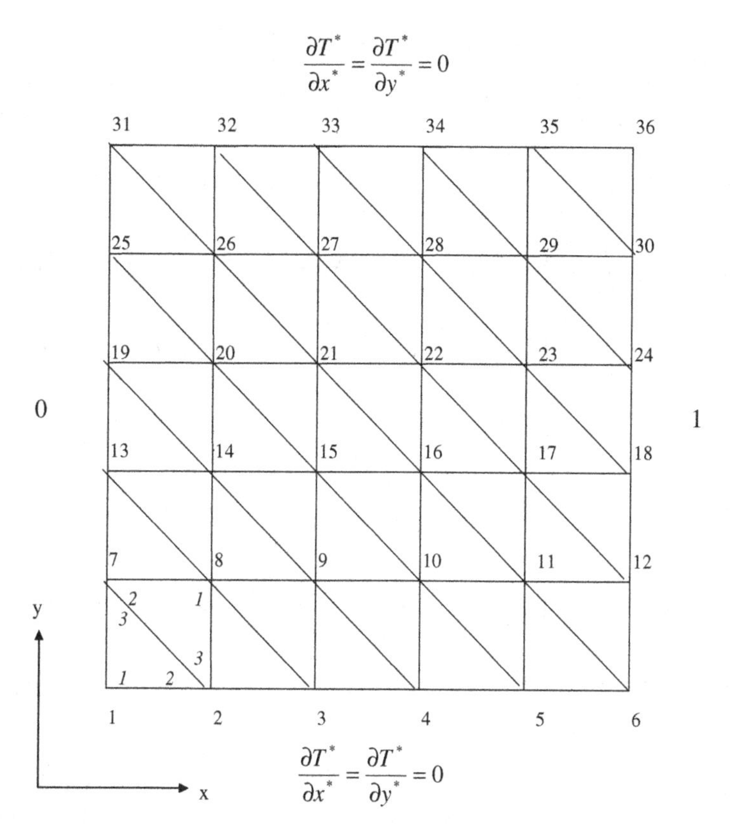

Fig. 4.14 Domain, boundary conditions, and mesh scheme for the solution of CDR equation using triangular elements.

Note that $x = \xi\ l_x$ and $y = \eta\ l_y$ and hence

$$\frac{d}{dx}(\Phi) = \frac{1}{l_x}\frac{d}{d\xi}(\Phi), \quad \frac{d}{dy}(\Phi) = \frac{1}{l_y}\frac{d}{d\eta}(\Phi), \quad dx = l_x d\xi, \quad dy = l_y d\eta.$$

After the transformation of Eq. (4.49) to the local coordinate system of (ξ, η)we have

$$\int_0^1 \int_0^{1-\xi} \left(\left(\frac{l_y}{l_x}T_\xi w_\xi + \frac{l_x}{l_y}T_\eta w_\eta\right) + l_x l_y D_a w T\right) d\eta d\xi = 0, \qquad (4.50)$$

and the members of elemental stiffness matrix is derived as

$$
\begin{cases}
E_{11} = \displaystyle\int_0^1 \int_0^{1-\xi} \left(w \left(\frac{l_y}{l_x} N_{1\xi} w_{1\xi} + \frac{l_x}{l_y} N_{1\eta} w_{1\eta} \right) + l_x l_y D_a N_1 w_1 \right) d\eta d\xi \\
E_{12} = \displaystyle\int_0^1 \int_0^{1-\xi} \left(\left(\frac{l_y}{l_x} N_{2\xi} w_{1\xi} + \frac{l_x}{l_y} N_{2\eta} w_{1\eta} \right) + l_x l_y D_a N_2 w_1 \right) d\eta d\xi \\
E_{13} = \displaystyle\int_0^1 \int_0^{1-\xi} \left(\left(\frac{l_y}{l_x} N_{3\xi} w_{1\xi} + \frac{l_x}{l_y} N_{3\eta} w_{1\eta} \right) + l_x l_y D_a N_3 w_1 \right) d\eta d\xi \\
E_{21} = \displaystyle\int_0^1 \int_0^{1-\xi} \left(\left(\frac{l_y}{l_x} N_{1\xi} w_{2\xi} + \frac{l_x}{l_y} N_{1\eta} w_{2\eta} \right) + l_x l_y D_a N_1 w_2 \right) d\eta d\xi \\
E_{22} = \displaystyle\int_0^1 \int_0^{1-\xi} \left(\left(\frac{l_y}{l_x} N_{2\xi} w_{2\xi} + \frac{l_x}{l_y} N_{2\eta} w_{2\eta} \right) + l_x l_y D_a N_2 w_2 \right) d\eta d\xi \\
E_{23} = \displaystyle\int_0^1 \int_0^{1-\xi} \left(\left(\frac{l_y}{l_x} N_{3\xi} w_{2\xi} + \frac{l_x}{l_y} N_{3\eta} w_{2\eta} \right) + l_x l_y D_a N_3 w_2 \right) d\eta d\xi \\
E_{31} = \displaystyle\int_0^1 \int_0^{1-\xi} \left(\left(\frac{l_y}{l_x} N_{1\xi} w_{3\xi} + \frac{l_x}{l_y} N_{1\eta} w_{3\eta} \right) + l_x l_y D_a N_1 w_3 \right) d\eta d\xi \\
E_{32} = \displaystyle\int_0^1 \int_0^{1-\xi} \left(\left(\frac{l_y}{l_x} N_{2\xi} w_{3\xi} + \frac{l_x}{l_y} N_{2\eta} w_{3\eta} \right) + l_x l_y D_a N_2 w_3 \right) d\eta d\xi \\
E_{33} = \displaystyle\int_0^1 \int_0^{1-\xi} \left(\left(\frac{l_y}{l_x} N_{3\xi} w_{3\xi} + \frac{l_x}{l_y} N_{3\eta} w_{3\eta} \right) + l_x l_y D_a N_3 w_3 \right) d\eta d\xi
\end{cases}
\quad (4.51)
$$

After the construction of the elemental stiffness equations the global set of stiffness equations are formed in the usual manner.

Chapters 1 and 2 describe in detail quadrature method that should be used to evaluate the integral terms in these matrices.

4.7 Multiscale Finite Element Methods for Time-Dependent Problems

Traditionally, in the finite element solution of time-dependent problems temporal discretizations are done separately from the spatial discretization. This is known as the decoupled formulation. In these formulations, spatial discretization is first performed by finite element method while temporal discretization is achieved using finite difference or finite element method (Lewis *et al.*, 2005). In another approach, known as the Taylor Galerkin method, Taylor series expansion of field unknown is used as a basis to replace temporal derivatives with spatial derivatives using the governing differential equation of a transient problem. Detailed discussion of these techniques can be found in many published sources (e.g. see Donea, 2003; Bell and Surana, 1996).

Alternative, fully coupled, space–time finite element discretization was first proposed by Fried (1969) and Oden (1969). In the coupled methods both spatial and temporal derivatives are simultaneously discretized. Therefore, a space–time mesh consisting of a large number of elements is used. Computational costs of such schemes are high and apparently offer no particular advantage over the decoupled schemes because geometrical flexibility of finite element method is irrelevant with respect to the time variable. To overcome the problems caused by excessive computational cost of coupled space–time discretizations, techniques such as layer-by-layer technique are used. Using this method in each time step a true initial value problem is posed and only one layer of elements in time is handled. For each time interval, therefore, initial conditions are taken as the end result of the previous interval and the procedure is repeated in a sequential manner.

Bubble function-enriched (multiscale) space–time discretizations resolve the problems associated with the use of meshes containing very large numbers of elements required in coupled techniques. Therefore, construction of cost-effective coupled space–time schemes becomes possible. Hughes and Stewart (1996) proposed a method for the development of multiscale variational schemes for time-dependent problems. Nassehi and Parvazinia (2009) proposed a practical multiscale procedure based on elements enriched by bubble functions. In the following section the basic concepts of multiscale space–time scheme are described.

4.7.1 *Multiscale space–time finite element discretization*

Let us consider the following approximation in terms of time-dependent shape functions for a transient field unknown given as

$$T(\tau) = N_1(\tau)T_{i1} + N_2(\tau)T_{i2}, \tag{4.52}$$

where T is the field unknown, τ is time variable, subscripts 1 and 2 denote two successive time levels and $N_1(\tau)$ and $N_2(\tau)$ are time-dependent shape functions.

To develop a multiscale scheme (here with respect to time variable) we need to enrich the shape functions with appropriate bubble functions. Using a local coordinate system $\tau(-1, +1)$ we have

$$N_1(\tau) = \frac{1}{2}(1+\tau) + \phi_1(\tau), \quad N_2(\tau) = \frac{1}{2}(1-\tau) + \phi_2(\tau), \tag{4.53}$$

where $\phi_1(\tau)$ and $\phi_2(\tau)$ are bubble functions. Time derivative of the field unknown is therefore written as

$$\frac{dT}{dt} = \frac{dN_1}{d\tau}T_{i1} + \frac{dN_2}{d\tau}T_{i2} = \left(\frac{1}{2} + \frac{d\phi_1}{d\tau}\right)T_{i1} + \left(-\frac{1}{2} + \frac{d\phi_2}{d\tau}\right)T_{i2}. \quad (4.54)$$

Using the matrix notation a system of first-order differential equations with respect to time can be written as

$$[M]\{\dot{T}\} + [G]\{T\} = \{B\}. \quad (4.55)$$

Substituting Eqs. (4.53) and (4.54) in Eq. (4.55) we have

$$\begin{aligned}[M]&\left\{\left(\frac{1}{2} + \frac{d\phi_1}{d\tau}\right)T_{i1} + \left(-\frac{1}{2} + \frac{d\phi_2}{d\tau}\right)T_{i2}\right\}\\ &+ [G]\left(N_1(\tau)T_{i1} + N_2(\tau)T_{i2}\right) = \{B\}.\end{aligned} \quad (4.56)$$

The selection of appropriate bubble functions makes it possible to use larger time steps in the solution of Eq. (4.56) and therefore to maintain computing economy.

4.7.2 *Multiscale space–time finite element modeling*

Let us consider the following problem defined in domain $\Omega \subset R^2$ as

$$\begin{cases} LT = f & \text{in } \Omega \\ T = 0 & \text{on } \Gamma \end{cases}, \quad (4.57)$$

where L is a time-dependent differential operator and f is a given source function defined on Ω (Hughes and Stewart, 1996).

Variational formulation which can be used to derive bubble functions for this problem is the same as the method used in previous sections. However, it must be noted that in a steady state problem the behavior of the field unknown is the same in all directions and hence elemental bubble functions have the same coefficient in all directions. In a transient problem the spatial and temporal behaviors are different by definition and using the same elemental bubble function is not possible. This problem is resolved by separating spatial bubble functions from temporal bubble functions. Therefore, the filed unknown (which in the present context will be represented in terms of "trial" or shape functions) and test functions (ν) used in the formulation of a variational statement for this problem can be written as

$$\begin{cases} T_b = T_{bs} + T_{bt} \\ v_b = v_{bs} + v_{bt} \end{cases}. \tag{4.58}$$

Therefore, the variational statement corresponding to Eq. (4.57) is written as

$$a(v_{bs} + v_{bt}, T_1) + a(v_{bs} + v_{bt}, T_{bs} + T_{bt}) = (v_{bs} + v_{bt}, f). \tag{4.59}$$

Equation (4.59) can be written as two sub-problems:

$$a(v_{bs}, T_1) + a(v_{bs}, T_{bs}) + a(v_{bs}, T_{bt}) = (v_{bs}, f), \tag{4.60}$$

$$a(v_{bt}, T_1) + a(v_{bt}, T_{bs}) + a(v_{bt}, T_{bt}) = (v_{bt}, f). \tag{4.61}$$

Therefore, STC required for the derivation of spatial and temporal bubble functions should be carried out separately.

4.7.3 *Two-dimensional space–time bubble functions for enrichment of bilinear elements*

Using bubble functions in x and t directions the bubble-enriched Lagrangian shape functions in local coordinate system $\xi(-1,+1), \tau(-1,+1)$can be written as

$$\begin{cases} N_1 = \dfrac{1}{4}(1-\xi)(1-\tau) + b\phi_\xi + b_t\phi_\tau \\ N_2 = \dfrac{1}{4}(1+\xi)(1-\tau) + b\phi_\xi + b_t\phi_\tau \\ N_3 = \dfrac{1}{4}(1+\xi)(1+\tau) + b\phi_\xi + b_t\phi_\tau \\ N_4 = \dfrac{1}{4}(1-\xi)(1+\tau) + b\phi_\xi + b_t\phi_\tau \end{cases}, \tag{4.62}$$

where ϕ_τ and $\phi_\xi$are temporal and spatial bubble functions, respectively. We now consider the following simple polynomials in terms of time variable:

$$\phi_\tau = \sum_{q=1}^{n} (1-\tau^2)^q, \tag{4.63}$$

$$\phi_\tau = \sum_{q=1}^{n} (1-\tau^{2q}). \tag{4.64}$$

For $q = 1$, we obtain a second-order temporal bubble function, for $q = 2$, we obtain a fourth-order temporal bubble function and so on.

4.7.4 *Elimination of the boundary integrals*

In a multidimensional case involving diffusion, the differential operator appearing in Eq. (4.57) will include a Laplacian term. Therefore, during the finite element solution we need to apply Green's theorem to reduce continuity requirement of selected elements (Chap. 1). This leads to the appearance of boundary integral terms which need to be considered and a technique for their elimination or calculation included in the finite element scheme. However, as can be seen in the following equation (which is based on transient CD equation) inclusion of temporal bubble function does not affect the Laplacian term:

$$\left(\frac{\partial T_h}{\partial t}, v\right) + (C\nabla T_h, v) + (D\nabla T_c, \nabla v) = (f, v). \tag{4.65}$$

4.7.5 *Transient CD problem*

We now consider the solution of transient CD equation using the described multiscale space–time method. Using vector notation, transient CD equation is given as

$$\rho c\left(\frac{\partial T}{\partial t} + \mathbf{v}.\nabla T\right) - k\nabla.\nabla T = f, \tag{4.66}$$

where T is the field variable, $\mathbf{v}$ is the velocity vector, k is diffusivity, ρ is density, c is heat capacity, and f is a source term. ∇ denotes the spatial gradient operator. Using the following dimensionless parameters

$$\begin{cases} T = T_0 T^* \\ x^* = \dfrac{x}{h},\ y^* = \dfrac{y}{h} \\ t^* = \dfrac{t}{t_0} \end{cases}, \tag{4.67}$$

where T_0 is reference value for the field variable, t_0 is a characteristic time interval, and h is a characteristic length (e.g. width of the domain), the transient CD equation is written in a dimensionless form as

$$\frac{\partial T^*}{\partial t^*} + C\nabla T^* - D\nabla.\nabla T^* = f^*. \tag{4.68}$$

Comparison of Eq. (4.68) with the original problem defined as Eq. (4.57) shows that the differential operator in this case can be written as

$$L = \frac{\partial}{\partial t^*} + C\nabla - D\Delta. \tag{4.69}$$

In which C and D are dimensionless convection and diffusion coefficients, respectively,

$$C = \frac{ut_0}{h}, \quad D = \frac{kt_0}{h^2\rho c}, \quad f^* = f\frac{t_0}{\rho c T_0}. \tag{4.70}$$

We now consider a two-dimensional transient CD equation given as

$$\frac{\partial T^*}{\partial t^*} + C\frac{\partial T^*}{\partial x^*} - D\frac{\partial^2 T^*}{\partial x^{*2}} = f^*. \tag{4.71}$$

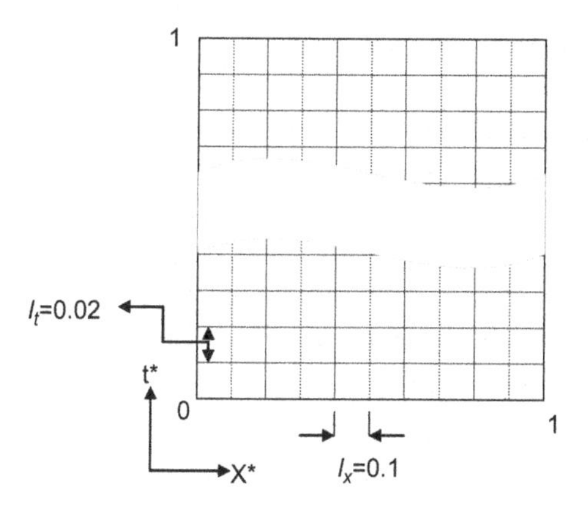

Fig. 4.15 Domain and meshes for the transient CD problem.

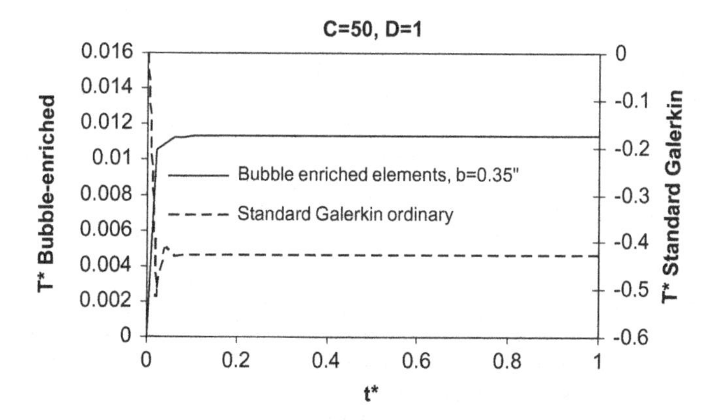

Fig. 4.16 Comparison of the results generated by the standard Galerkin and multiscale schemes for the transient CD equation (D=1 and C=50).

Equation (4.71) is solved subject to the following dimensionless boundary conditions in a rectangular domain

$$\begin{cases} T^* = 0 & \text{for } x^* = 0,\ 0 \le t^* \le 1 \\ T^* = 1 & \text{for } x^* = 1,\ 0 \le t^* \le 1 \\ T^* = 0 & \text{for } t^* = 0,\ 0 \le x^* \le 1 \end{cases} . \tag{4.72}$$

It is assumed that the spatial boundaries are fixed during the time that the problem is being solved and, therefore, any temporal discretization can be done over the entire time domain. Using the described space–time bubble functions, the normal procedure of the finite element method is carried out. The results obtained for the rectangular domain shown in Fig. 4.15 at $x^* = 0.9$ for values of $D = 1$ and $C = 50$ at different times using both the standard Galerkin and the described multiscale schemes are shown in Fig. 4.16. As shown here the result produced by the standard scheme is unstable. In contrast the multiscale scheme has generated theoretically expected output. Spatial enrichment is done using second-order bubble function and temporal enrichment using tenth-order bubble function. The bubble coefficient is set to be $b = 0.35$ for both temporal and spatial enrichments.

For further explanations of transient multiscale schemes see Nassehi and Parvazinia (2009).

References

[1] Bell, B.C. and Surana, K.S., A space–time coupled p-version least squares finite element formulation for unsteady two-dimensional Navier–Stokes equations, *Int. J. Numer. Meth. Eng.*, 1996; 39; 2593–2618.

[2] Brodkey, R.S. and Hershey, H.C., *Transport Phenomena*, McGraw Hill, 1988.

[3] Donea, J. and Taylor, A., Galerkin method for convective transport problems, *Int. J. Numer. Meth. Eng.*, 1984; 20; 101–120.

[4] Donea, J., Roig, B. and Huerta, A., High order accurate time stepping schemes for convection–diffusion equation, *Comput. Meth. Appl. Mech. Eng.*, 2000; 182; 249–275.

[5] Donea, J. and Huerta, A., *Finite Element Methods for Flow Problems*, John Wiley & Sons, Chichester, 2003.

[6] Franca, L.P., Hughes, T.J.R. and Stenberg, R., Stabilized finite element methods for the Stokes problem, in Nicolaides, R.A. and Gunzberger, M.D. (Eds.), *Incompressible Fluid Dynamics — Trends and Advances*, Cambridge University Press, Cambridge, 1993.

[7] Franca, L.P., Farhat, C., Macedo, A.P. and Lesoinne, M., Residual free bubbles for Helemholtz equation, *Comput. Meth. Appl. Mech. Eng.*, 1997; 40; 4003–4009.

[8] Franca, L.P. and Russo, A., Deriving upwinding, mass lumping and selective reduced integration by residual free bubbles, *Appl. Math. Lett.*, 1996; 9; 83–88.
[9] Franca, L.P. and Russo, A., Unlocking with residual-free bubbles, *Comput. Meth. Appl. Mech. Eng.*, 1997a; 142; 361–364.
[10] Franca, L.P. and Russo, A., Mass lumping emanating from residual free bubbles, *Comput. Meth. Appl. Mech. Eng.*, 1997b; 142; 353–360.
[11] Franca, L.P. and Farhat, C., Bubble functions prompt unusual stabilized finite element methods, *Comput. Meth. Appl. Mech. Eng.*, 1995; 123; 299–308.
[12] Fried, I., Finite-element analysis of time-dependent phenomena, *AIAA J*, 1969; 7; 1170–1172.
[13] Hughes, T.J.R. and Hulbert, G.M., Space–time finite element methods for elastodynamics: Fomulations and error estimates, *Comput. Meth. Appl. Mech. Eng.*, 1998; 66; 339–363.
[14] Hughes, T.J.R. and Stewart, J., A space time formulation for multiscale phenomena, *J. Comput. Appl. Math.*, 1996; 74; 217–229.
[15] Hughes, T.J.R., Multiscale phenomena, Green's functions, the Dirichlet-to-Neumann formulation, subgrid scale models, bubbles and the origins of stabilized methods, *Comput. Meth. Appl. Mech. Eng.*, 1995; 127; 381–401.
[16] Lewis, R.W., Nithiarasu, P. and Seetharamu, K.N., *Fundamentals of the Finite Element Method for Heat and Fluid Flow*, John Wiley & Sons, Chichester, 2005.
[17] Nassehi, V., *Practical Aspects of Finite Element Modelling of Polymer Processing*, John Wiley & Sons, Chichester, 2002.
[18] Nassehi, V., Numerical modelling of tidal dynamics in a one-dimensional branching estuary, PhD thesis, University of Wales, Swansea, UK, 1981.
[19] Nassehi, V. and Parvazinia, M., A multiscale finite element space–time discretization method for transient transport phenomena using bubble functions, *Finite Elements in Analysis and Design*, 2009; 45; 315–323.
[20] Oden, J.T., A general theory of finite element, Part I and II, *Int. J. Numer. Meth. Eng.*, 1969; 1; 205–221, 247–259.
[21] Parvazinia, M., Nassehi, V. and Wakeman, R.J, Multiscale finite element modelling of laminar steady flow through highly permeable porous media, *Chem. Eng. Sci.*, 2006; 61; 586–596.
[22] Parvazinia, M. and Nassehi, V., Multiscale finite element modeling of diffusion–reaction equation using bubble functions with bilinear and triangular elements, *Comput. Meth. Appl. Mech. Eng.*, 2007; 196; 1095–1107.
[23] Parvazinia, M. and Nassehi, M., Using bubble functions in the multi-scale finite element modeling of the convection–diffusion–reaction equation, *Int. Commun. Heat Mass Transfer*, 2010; 37; 125–130.

CHAPTER 5

Application of Multiscale Finite Element Schemes to Fluid Flow Problems

In this chapter, we discuss the multiscale finite element solution of a number of flow problems. The selected examples represent cases in which multiscale behavior may be present and, therefore, finite element schemes based on ordinary types of discretizations may need excessively refined mesh to generate reliable solutions.

5.1 Two-Dimensional and Axisymmetric Flow Regimes — Governing Equations

(a) Free flow

Using a two-dimensional Cartesian (x, y) coordinate system the components of equation of motion, based on conservation of momentum, for the laminar flow of an incompressible Newtonian fluid with constant density are expressed as

x-component

$$\rho\left(\frac{\partial v_x}{\partial t}+v_x\frac{\partial v_x}{\partial x}+v_y\frac{\partial v_x}{\partial y}\right)=-\frac{\partial p}{\partial x}+\mu\left(\frac{\partial^2 v_x}{\partial x^2}+\frac{\partial^2 v_x}{\partial y^2}\right)+\rho g_x, \tag{5.1}$$

y-component

$$\rho\left(\frac{\partial v_y}{\partial t}+v_x\frac{\partial v_y}{\partial x}+v_y\frac{\partial v_y}{\partial y}+\right)=-\frac{\partial p}{\partial y}+\mu\left(\frac{\partial^2 v_y}{\partial x^2}+\frac{\partial^2 v_y}{\partial y^2}\right)+\rho g_y, \tag{5.2}$$

where v_x and v_y are components of velocity, p is pressure, ρ and μ are fluid density and viscosity respectively, and g_x and g_y are components of body force acting on the flow regime. The terms in the left-hand side of Eqs. (5.1) and (5.2) are inertia terms and represent local and convective acceleration in a flow field. The first and second terms in the right-hand side of these equations represent surface forces due to pressure gradient and viscous stress acting on the fluid.

Similarly, for axisymmetric flow the components of equation of motion in terms of radial and axial (r, z) coordinates for an incompressible Newtonian fluid with constant density are given as

r-component

$$\rho\left(\frac{\partial v_r}{\partial t}+v_r\frac{\partial v_r}{\partial r}+v_z\frac{\partial v_r}{\partial z}\right)=-\frac{\partial p}{\partial r}+\mu\left(\frac{\partial}{\partial r}\left(\frac{1}{r}\frac{\partial(rv_r)}{\partial r}\right)+\frac{\partial^2 v_r}{\partial z^2}\right)+\rho g_r, \tag{5.3}$$

z-component

$$\rho\left(\frac{\partial v_z}{\partial t}+v_r\frac{\partial v_z}{\partial r}+v_z\frac{\partial v_z}{\partial z}\right)=-\frac{\partial p}{\partial z}+\mu\left(\frac{1}{r}\frac{\partial}{\partial r}\left(r\frac{\partial v_z}{\partial r}\right)+\frac{\partial^2 v_z}{\partial z^2}\right)+\rho g_z. \tag{5.4}$$

It should be noted that in an axisymmetric flow all dependent variables remain unchanged in the circumferential (θ) direction, and the integration of the governing equation with respect to this coordinate variable results in constant value of 2π being multiplied by the remaining components of the equation of motion and can be dropped. Equations. (5.3) and (5.4), therefore, represent a three-dimensional flow regime.

(b) Porous flow

A wide range of mathematical models have been used to represent porous flow regimes. These range from equations based on Darcy's law which provides a linear relation between flux and pressure gradients in a low permeability porous medium to more general equations which involve convection, inertia and viscous effects. Components of the simplest equation representing porous flow are therefore written as

$$\begin{cases} v_x=-\dfrac{K}{\mu}\dfrac{\partial p}{\partial x} \\ v_y=-\dfrac{K}{\mu}\dfrac{\partial p}{\partial y} \end{cases}, \tag{5.5}$$

where K is a permeability coefficient. Darcy's equation (Eq. (5.5)) does not include any inertia terms and it represents a situation in which stress is entirely borne by the matrix of a porous flow field (i.e. the medium) and fluid itself is not affected by internal stress forces.

The free and porous flow regimes described so far do not normally exhibit multiscale behavior. Free flow equations presented here (known as the Navier–Stokes equations) characterize regimes in which beyond the entry region in a Newtonian fluid flow domain a velocity field with a parabolic profile develops. Porous flow regimes described by the Darcy equation are characterized by a flat velocity profile (plug flow). In a significant number of instances, however, a flow regime exhibits a combination of free and porous flow systems. For example, in porous flows through domains of high permeability, the medium (i.e. the porous matrix) carries only a part of stress imposed on the fluid and the fluid itself also carries some stress. This effect is especially important near solid walls of such a domain. Consequently, the profile of the velocity field in a porous flow through a highly permeable medium is similar to a free flow near the walls, and to a Darcy flow inside the domain. Depending on the permeability of the medium, the layer near the walls, where the velocity profile is no longer flat, can be extremely narrow and hence a multiscale situation develops. To take into account the described effects, various combinations of free and porous flow models have been suggested. The most commonly used form is the Brinkman model (Nield and Bejan, 1992), which in a two-dimensional domain can be expressed as

x-component of the Brinkman equation

$$-\frac{\partial p}{\partial x} + \mu_e \left(\frac{\partial^2 v_x}{\partial x^2} + \frac{\partial^2 v_x}{\partial y^2} \right) - \frac{\mu}{K} v_x = 0, \tag{5.6}$$

y-component of the Brinkman equation

$$-\frac{\partial p}{\partial y} + \mu_e \left(\frac{\partial^2 v_y}{\partial x^2} + \frac{\partial^2 v_y}{\partial y^2} \right) - \frac{\mu}{K} v_y = 0, \tag{5.7}$$

where μ is fluid viscosity, K is the domain permeability, and μ_e is an effective viscosity that theoretically takes into account the stress borne by the fluid as it flows through a porous medium. Experimental measurement of μ_e is not a trivial matter, if not impossible. Therefore, in the rest of this chapter μ_e is set to be equal to the fluid viscosity μ (this is a commonly made assumption in using the Brinkman equation, e.g. see Hsu and Cheng, 1985; Kaviany, 1986; Allan and Hamdan, 2002).

The described flow equations are always solved in conjunction with the continuity equation, which is the expression of mass conservation. In two-dimensional cases this equation for an incompressible fluid is expressed as

$$\frac{\partial v_x}{\partial x} + \frac{\partial v_y}{\partial y} = 0. \tag{5.8}$$

Axisymmetric continuity equation for an incompressible fluid flow is given as

$$\frac{\partial v_r}{\partial r} + \frac{v_r}{r} + \frac{\partial v_z}{\partial z} = 0. \tag{5.9}$$

In cases where flow regime is not isothermal, or heat transport by a flowing fluid is being modeled, the flow and continuity equations should be solved together with an energy equation. Equation of energy in a Cartesian coordinate system for a Newtonian fluid with constant physical coefficients is written as

$$\begin{aligned}\rho C_p\left(\frac{\partial T}{\partial t} + v_x\frac{\partial T}{\partial x} + v_y\frac{\partial T}{\partial y}\right) &= k\left(\frac{\partial^2 T}{\partial x^2} + \frac{\partial^2 T}{\partial y^2}\right)\\ &\quad + 2\mu\left\{\left(\frac{\partial v_x}{\partial x}\right)^2 + \left(\frac{\partial v_y}{\partial y}\right)^2\right\}\\ &\quad + \mu\left(\frac{\partial v_x}{\partial y} + \frac{\partial v_y}{\partial x}\right)^2,\end{aligned} \tag{5.10}$$

where T is temperature, k is conductivity, and C_p is specific heat capacity under constant pressure. Equation of energy in an axisymmetric system, for a Newtonian fluid with constant physical coefficients is also given as

$$\begin{aligned}\rho C_p\left(\frac{\partial T}{\partial t} + v_r\frac{\partial T}{\partial r} + v_z\frac{\partial T}{\partial z}\right) &= k\left(\frac{1}{r}\frac{\partial}{\partial r}\left(r\frac{\partial T}{\partial r}\right) + \frac{\partial^2 T}{\partial z^2}\right)\\ &\quad + 2\mu\left\{\left(\frac{\partial v_r}{\partial r}\right)^2 + \left(\frac{v_r}{r}\right)^2 + \left(\frac{\partial v_z}{\partial z}\right)^2\right\}\\ &\quad + \mu\left(\frac{\partial v_z}{\partial r} + \frac{\partial v_r}{\partial z}\right)^2.\end{aligned} \tag{5.11}$$

5.2 Modeling of Isothermal Brinkman Flow of a Newtonian Fluid in a Two-Dimensional Domain Using Multiscale Finite Element Schemes

Consider a high permeability porous duct (schematically represented as Fig. 5.1). Equations. (5.6)–(5.8) provide the governing equations of a mathematical model describing the Brinkman flow of a Newtonian fluid within

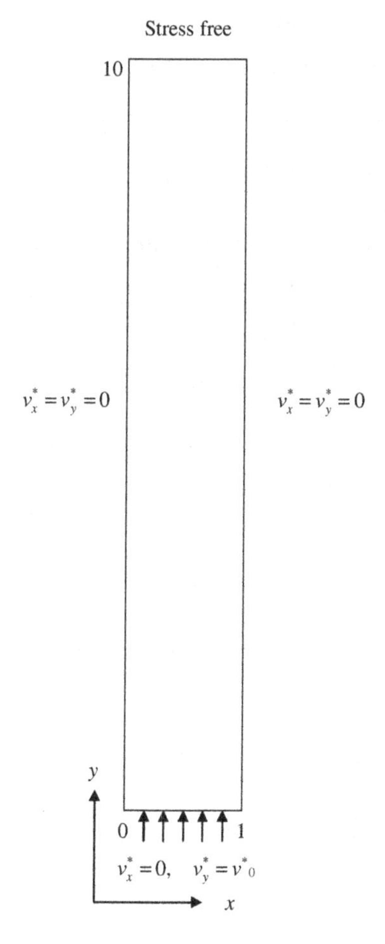

Fig. 5.1 Flow domain and boundary conditions.

this domain. The prescribed boundary conditions are

(I) Inlet to the domain
At the inlet a plug flow condition is imposed (this is in accordance with majority of engineering flow conditions). Therefore,

$$v_x = 0, v_y = v_0 \quad \text{for } 0 < x < h_x \text{ and } y = 0, \tag{5.12}$$

where h_x is the gap width in the rectangular domain.

(II) Impermeable (solid) walls

$$v_x = 0, \quad v_y = 0 \quad \text{for } x = 0 \text{ and } 0 \leq y < h_y,$$
$$v_x = 0, \quad v_y = 0 \quad \text{for } x = h_x \text{ and } 0 \leq y < h_y. \tag{5.13}$$

(III) Exit

At the outlet a stress-free condition is used, therefore, both shear and normal components of the surface forces are set to zero.

$$\tau_{yy}|_{\text{exit}} = -2\mu \frac{\partial v_y}{\partial y} = 0 \quad \text{for } y = h_y \text{ and } 0 \leq x \leq h_x, \tag{5.14}$$

$$\tau_{yx}|_{\text{exit}} = \tau_{xy}|_{\text{exit}} = -\mu \left(\frac{\partial v_x}{\partial y} + \frac{\partial v_y}{\partial x} \right)$$
$$= 0 \quad \text{for } y = h_y \text{ and } 0 \leq x \leq h_x, \tag{5.15}$$

$$\tau_{xx}|_{\text{exit}} = -2\mu \frac{\partial v_x}{\partial x} = 0 \quad \text{for } y = h_y \text{ and } 0 \leq x \leq h_x. \tag{5.16}$$

The use of "stress-free" instead of "developed flow" conditions provides more general exit boundary conditions enabling the simulation of realistic situations where the complete development of the flow regime cannot be guaranteed.

To preserve the consistency of the numerical solutions, we use the following dimensionless variables as initially introduced by Parvazinia *et al.* (2006a)

$$y^* = \frac{y}{h}, \quad x^* = \frac{x}{h}, \quad v_x^* = \frac{v_x \mu}{\rho g h^2}, \quad v_y^* = \frac{v_y \mu}{\rho g h^2}, \quad p^* = \frac{p}{\rho g h},$$
$$\tau_{xx}^* = \frac{\tau_{xx}}{\rho g h}, \quad \tau_{yy}^* = \frac{\tau_{yy}}{\rho g h}, \quad \tau_{yx}^* = \frac{\tau_{yx}}{\rho g h}, \quad \tau_{xy}^* = \frac{\tau_{xy}}{\rho g h}, \tag{5.17}$$

where ρ is fluid density, p is the pressure, and g is acceleration due to gravity. Substituting defined dimensionless variables in the governing Eqs. (5.6)–(5.8) and boundary conditions (5.12)–(5.16), the following dimensionless equations are obtained:

$$\frac{\partial v_x^*}{\partial x^*} + \frac{\partial v_y^*}{\partial y^*} = 0, \tag{5.18}$$

$$-\frac{\partial p^*}{\partial x^*} - \frac{1}{Da} v_x^* + \left(\frac{\partial^2 v_x^*}{\partial x^{*2}} + \frac{\partial^2 v_x^*}{\partial y^{*2}} \right) = 0, \tag{5.19}$$

$$-\frac{\partial p^*}{\partial y^*} - \frac{1}{Da} v_y^* + \left(\frac{\partial^2 v_y^*}{\partial x^{*2}} + \frac{\partial^2 v_y^*}{\partial y^{*2}} \right) = 0, \tag{5.20}$$

where D_a is the Darcy parameter defined as

$$D_a = \frac{K}{h^2}.$$

We choose $h_y = 10h_x$. The corresponding dimensionless boundary conditions are expressed as

(I) Inlet

$$v_x^* = 0, \quad v_y^* = v_0^* \quad \text{for } y^* = 0 \text{ and } 0 < x^* < 1, \tag{5.21}$$

here v_0^* is the inlet velocity of the plug flow considered.

(II) Impermeable walls

$$\begin{aligned} v_x^* = v_y^* = 0 \quad &\text{for } x^* = 0 \text{ and } 0 \leq y^* < 10, \\ v_x^* = v_y^* = 0 \quad &\text{for } x^* = 1 \text{ and } 0 \leq y^* < 10. \end{aligned} \tag{5.22}$$

(III) Exit

Stress-free conditions expressed in the dimensionless form are imposed.

$$\tau_{yy}^* = -2\frac{\partial v_y^*}{\partial y^*} = 0 \quad \text{for } y^* = 10 \text{ and } 0 \leq x^* \leq 1, \tag{5.23}$$

$$\begin{aligned} \tau_{yx}^* = \tau_{xy}^* &= -\left(\frac{\partial v_x^*}{\partial y^*} + \frac{\partial v_y^*}{\partial x^*}\right) \\ &= 0 \quad \text{for } y^* = 10 \text{ and } 0 \leq x^* \leq 1, \end{aligned} \tag{5.24}$$

$$\tau_{xx}^* = -2\frac{\partial v_x^*}{\partial x^*} = 0 \quad \text{for } y^* = 10 \text{ and } 0 \leq x^* \leq 1. \tag{5.25}$$

5.2.1 *Continuous penalty scheme*

There are a variety of differently weighted residual finite element techniques that can be used to solve the governing equations of various flow regimes. Both ordinary and multiscale schemes can be formulated using these techniques as the main solution strategy. In this section, we briefly describe the continuous penalty technique (Pittman, 1989). This technique is, in essence, similar to the "Lagrange Multiplier Method" used for the solution of differential equations subject to a constraint. Here the continuity equation (i.e. the incompressibility condition) is regarded as a constraint for the equation of motion. Therefore, instead of solving the governing flow equations as a system of three PDEs the pressure in the components of the equation of motion is replaced by a multiplier (called penalty parameter) times the continuity equation. This gives a more compact set of working equations with components of the velocity as the remaining unknowns. Additionally, elimination of the pressure from the equation of motion automatically satisfies the basic numerical stability condition for the simulation of incompressible flows, known as the LBB criteria. The mathematical theory underpinning the formulation of LBB criterion is somewhat obscure (Reddy, 1986). However, it can be readily observed that the absence of a pressure term in the incompressible continuity equation makes the possibility of a mismatch between approximations used to satisfy the equations of motion and continuity almost inevitable in any numerical solution of a system of PDEs with velocity and pressure as the prime unknowns.

As our aim is to use bubble function-enriched elements, we note that the LBB condition remains satisfied using enriched bilinear elements (Bai, 1997).

In the continuous penalty technique to eliminate the pressure from the governing equations, pressure is expressed in terms of the incompressibility condition as

$$p = -\lambda \left(\frac{\partial v_x}{\partial x} + \frac{\partial v_y}{\partial y} \right). \tag{5.26}$$

It can be shown that by using a sufficiently large penalty parameter (λ) the incompressibility condition is satisfied (Nassehi, 2002). However, insertion of a very large number into the system of governing equations upsets the relative magnitude of the terms in the equation of motion making the system ill-conditioned. To minimize such effects and to maintain the consistency of the terms in the working equations of the scheme, it is advantageous to make the penalty parameter proportional to the fluid viscosity as

$$\lambda = \mu \lambda_0, \tag{5.27}$$

λ_0 is dimensionless and large enough to generate a sufficiently large λ. Substituting Eq. (5.27) into Eq. (5.26) the dimensionless form of Eq. (5.26) is found as

$$p^* = -\lambda_0 \left(\frac{\partial v_x^*}{\partial x^*} + \frac{\partial v_y^*}{\partial y^*} \right). \tag{5.28}$$

Substituting Eq. (5.28) in the dimensionless form of Eqs. (5.19) and (5.20) we get

$$-\frac{\partial}{\partial x^*}\left[-\lambda_0\left(\frac{\partial v_x^*}{\partial x^*}+\frac{\partial v_y^*}{\partial y^*}\right)\right]-\frac{1}{D_a}v_x^*+\left(\frac{\partial^2 v_x^*}{\partial x^{*2}}+\frac{\partial^2 v_x^*}{\partial y^{*2}}\right)=0, \quad (5.29)$$

$$-\frac{\partial}{\partial y^*}\left[-\lambda_0\left(\frac{\partial v_x^*}{\partial x^*}+\frac{\partial v_y^*}{\partial y^*}\right)\right]-\frac{1}{D_a}v_y^*+\left(\frac{\partial^2 v_y^*}{\partial x^{*2}}+\frac{\partial^2 v_y^*}{\partial y^{*2}}\right)=0. \quad (5.30)$$

After the discretization of the solution domain into a computational mesh, consisting of finite elements of predetermined geometrical shapes, the prime unknowns in the governing equations are replaced by approximate forms defined within the selected finite elements. In the weighted residual finite element scheme, these unknowns are replaced by trial function representations, which in the context of a discretized domain are given by low order interpolation polynomials, N_j (Zienkiewicz and Taylor, 1994):

$$v_x^* = \sum_{j=1}^{n} N_j v_{xj}^*, \quad (5.31)$$

$$v_y^* = \sum_{j=1}^{n} N_j v_{yj}^*, \quad (5.32)$$

where n is total number of nodes in an element and, v_{xj}^* and v_{yj}^* are nodal values of unknowns (i.e. values at the sampling points in an element). Therefore, Eqs. (5.31) and (5.32) provide approximate values for unknowns within an element via interpolation using their nodal values. Substitution of approximate values for the velocity components from Eqs. (5.31) and (5.32) into Eqs. (5.29) and (5.30), leads to the appearance of residual statements. These statements are then multiplied by appropriate weight functions (w_i) and integrated over an elemental domain

$$\int_{\Omega_e} W_i\left[\frac{\partial}{\partial x^*}\lambda_0\left(\frac{\partial\sum_{j=1}^{n}N_j v_{xj}^*}{\partial x^*}+\frac{\partial\sum_{j=1}^{n}N_j v_{yj}^*}{\partial y^*}\right)-\frac{1}{Da}\sum_{J=1}^{n}N_j v_{xj}^*\right.$$
$$\left.+\left(\frac{\partial^2\sum_{j=1}^{n}N_j v_{xj}^*}{\partial x^{*2}}+\frac{\partial^2\sum_{j=1}^{n}N_j v_{xj}^*}{\partial y^{*2}}\right)\right]dx^*dy^*=0 \quad (5.33)$$

$$\int_{\Omega_e} W_i \left[\frac{\partial}{\partial y^*} \lambda_0 \left(\frac{\partial \sum_{j=1}^{n} N_j v_{xj}^*}{\partial x^*} + \frac{\partial \sum_{j=1}^{n} N_j v_{yj}^*}{\partial y^*} \right) - \frac{1}{Da} \sum_{J=1}^{n} N_j v_{yj}^* \right.$$

$$\left. + \left(\frac{\partial^2 \sum_{j=1}^{n} N_j v_{yj}^*}{\partial x^{*2}} + \frac{\partial^2 \sum_{j=1}^{n} N_j v_{yj}^*}{\partial y^{*2}} \right) \right] dx^* dy^* = 0, \tag{5.34}$$

where W_i is a weight function. In the standard Galerkin method the weight functions used to form weighted residual statements are selected to be identical to the shape functions (Chap. 1). However in the multi-scale model considered here, the shape functions (i.e. N_j) consist of a part based on Lagrangian shape function ψ_j associated with any selected element type and a bubble function part (Chap. 3). In Eqs. (5.33) and (5.34) the weight functions are chosen to be identical to the Lagrangian part of the shape functions. The second-order differentials in Eqs. (5.33) and (5.34) are reduced by the application of Green's theorem (i.e. generalized form of integration by parts). This leads to the appearance of boundary integral (flux) terms along the exterior boundaries of finite elements. For each interpolation function an identical weight function can be used to generate weighted residual equations such as Eqs. (5.33) and (5.34). Therefore, corresponding to a total of n interpolation functions (determined by the number of nodes in the selected type of element for domain discretization), n equations are generated and a system of $n \times n$ equations is constructed. Using matrix notation this system is written as

$$\begin{bmatrix} A_{ij}^{11} & A_{ij}^{12} \\ A_{ij}^{21} & A_{ij}^{22} \end{bmatrix} \begin{Bmatrix} v_{xj}^* \\ v_{yj}^* \end{Bmatrix} = \begin{Bmatrix} B_j^1 \\ B_j^2 \end{Bmatrix}, \tag{5.35}$$

where

$$A_{ij}^{11} = \int_{\Omega_e} \left[(\lambda_0 + 1) \left(\frac{\partial W_i}{\partial x^*} \frac{\partial N_j}{\partial x^*} \right) + \frac{\partial W_i}{\partial y^*} \frac{\partial N_j}{\partial y^*} \right.$$

$$\left. + \frac{1}{D} W_i N_j \right] dx^* dy^*, \tag{5.36}$$

$$A_{ij}^{12} = \int_{\Omega_e} \lambda_0 \frac{\partial W_i}{\partial x^*} \frac{\partial N_j}{\partial y^*} dx^* dy^*, \tag{5.37}$$

$$A_{ij}^{21} = \int_{\Omega_e} \lambda_0 \frac{\partial W_i}{\partial y^*} \frac{\partial N_j}{\partial x^*} dx^* dy^*, \tag{5.38}$$

$$A_{ij}^{22} = \int_{\Omega_e} \left[(\lambda_0 + 1) \left(\frac{\partial W_i}{\partial y^*} \frac{\partial N_j}{\partial y^*} \right) + \frac{\partial W_i}{\partial x^*} \frac{\partial N_j}{\partial x^*} + \frac{1}{Da} W_i N_j \right] dx^* dy^*, \tag{5.39}$$

$$B_j^1 = \int_{\Gamma_e} W_i \left\{ \left[\lambda_0 \left(\frac{\partial v_x^{*e}}{\partial x^*} + \frac{\partial v_y^{*e}}{\partial y^*} \right) + \frac{\partial v_x^{*e}}{\partial x^*} \right] n_x + \left(\frac{\partial v_x^{*e}}{\partial y^*} \right) n_y \right\} d\Gamma_e, \tag{5.40}$$

$$B_j^2 = \int_{\Gamma_e} W_i \left\{ \left(\frac{\partial v_y^{*e}}{\partial x^*} \right) n_x + \left[\lambda_0 \left(\frac{\partial v_x^{*e}}{\partial x^*} + \frac{\partial v_y^{*e}}{\partial y^*} \right) + \frac{\partial v_y^{*e}}{\partial y^*} \right] n_y \right\} d\Gamma_e. \tag{5.41}$$

Note that by using an elemental coordinate system and isoparametric mapping (Chap. 1) instead of the global coordinates, an elemental matrix equation which remains uniform over all of the elements in a discretized domain is derived. In practice, terms of Eq. (5.35) (i.e. Eqs. (5.36)–(5.41)) are transformed into a master element using isoparametric mapping. In addition, a natural coordinate system such as $-1 \leq \xi, \eta \leq +1$ can be used within the master element to enable the evaluation of all integrals within its domain by Gauss quadrature method (Gerald and Wheatley, 1984).

Repeated application of the above procedure to each element in the computational mesh leads to the construction of elemental weighted residual equations. Subsequent assembly of these equations over the common nodes between elements provides a system of global algebraic equations. The flux terms along all interior element boundaries cancel out each other leaving only the boundary integrals along the exterior boundaries of the solution domain. These terms should then be treated via the imposition of boundary conditions to obtain a determinate set of equations. At the inlet the velocity components are known and hence equations corresponding to the nodes along this line are dropped from the global set. As Eqs. (5.40) and (5.41) show, through the imposition of stress-free conditions at exit (Eqs. (5.23)–(5.25)) the line integrals along the exit boundary give a uniform value of zero. At the no-slip solid walls known velocity components (i.e. $v_x^* = v_y^* = 0$) are imposed and terms involving boundary integrals are dropped.

After the imposition of all boundary conditions into the assembled set of working equations, a determinate set is derived. The solution of global set of equations, can be based on direct or iterative techniques (Nassehi, 2002). A computationally efficient version of the Gaussian elimination method which relies on reduction of the global system to upper triangular form by an advancing front (Irons, 1970) is the most commonly used technique in most types of fluid flow simulations.

Calculation of pressure in the continuous penalty scheme is based on Eq. (5.28) and starts after obtaining the velocity field. The method used is called variational recovery (see Nassehi, 2002).

5.2.2 *Bubble-enriched shape functions used in conjunction with the continuous penalty method*

The procedure for the derivation of bubble functions used to solve the Brinkman equation by the continuous penalty finite element scheme is similar to those explained for both residual free bubble (RFB) and static condensation (STC) methods in Chap. 3. However, in order to develop a practical procedure in this case we need to assume a constant pressure drop which can later be eliminated during the calculations to obtain a homogeneous equation. To develop an appropriate form for the RFB method, let us consider the following bubble-enriched shape functions corresponding to a four-node rectangular element:

$$\begin{cases} N_1 = \dfrac{1}{4}(1-\xi)(1-\eta) + b(1-\xi^2)(1-\eta^2)(3-\xi)(3-\eta) \\ N_2 = \dfrac{1}{4}(1+\xi)(1-\eta) + b(1-\xi^2)(1-\eta^2)(3+\xi)(3-\eta) \\ N_3 = \dfrac{1}{4}(1+\xi)(1+\eta) + b(1-\xi^2)(1-\eta^2)(3+\xi)(3+\eta) \\ N_4 = \dfrac{1}{4}(1-\xi)(1+\eta) + b(1-\xi^2)(1-\eta^2)(3-\xi)(3+\eta) \end{cases}, \tag{5.42}$$

where

$$b = -\frac{1}{8}\left(1 + \frac{6D_a}{l^2}\right)^{-1}.$$

5.2.3 *Solved example — unidirectional flow*

To evaluate the accuracy of the numerical solutions obtained using bubble-enriched elements for the Brinkman equation they are compared with an analytical solution. The dimensionless Brinkman equation in one dimension corresponding to a constant pressure drop can be written as (Parvazinia *et al.*, 2006a)

$$\begin{cases} \dfrac{d^2v^*}{dx^{*2}} - \dfrac{1}{D_a}v^* + p_d^* = 0 \\ v^* = 0 \quad \text{at } x^* = 0 \\ v^* = 0 \quad \text{at } x^* = 1 \end{cases}, \tag{5.43}$$

where

$$p_d^* = -\frac{\partial P^*}{\partial y^*}. \tag{5.44}$$

Solution of the above equation gives

$$v^* = \frac{p_d^* D_a(e^{-\alpha} - 1)}{(e^{\alpha} - e^{-\alpha})}(e^{\alpha x^*} - e^{-\alpha x^*}) + p_d^* D_a(1 - e^{-\alpha x^*}), \tag{5.45}$$

where $\alpha = \sqrt{\frac{1}{D_a}}$.

To calculate the corresponding pressure field we assume an average velocity equal to plug flow velocity at the inlet. In the present domain it is written as

$$\int_0^1 v^* dx^* = v_0^*. \tag{5.46}$$

Solution of Eq. (5.43) gives

$$p_d^* = \frac{1}{D_a} v_0^* \left(1 + \frac{1}{\alpha}\left(e^{-\alpha} - 1\right)\left(1 + \frac{D_a(e^{\alpha} - e^{-\alpha} - 2)}{e^{\alpha} - e^{-\alpha}}\right)\right)^{-1}. \tag{5.47}$$

Excess pressure loss due to the entrance region can be neglected (Kaviany, 1986).

In Fig. 5.2 velocity profile at a sample cross-section inside the porous flow domain, generated by the continuous penalty finite element scheme based on ordinary bilinear elements, is given and compared with the results obtained using the same scheme based on bubble function-enriched elements. As shown in this figure, results obtained using the ordinary elements is oscillatory and inaccurate. Bubble function-enriched elements, on the other hand, generate significantly better results. It should be noted that in the unidirectional flow simulated in this example, computational results can be compared with the analytical solution obtained using one-dimensional governing equations (i.e. Eq. (5.45)). In this example, $D_a = 4 \times 10^{-4}$ and $v_0^* = 0.2$.

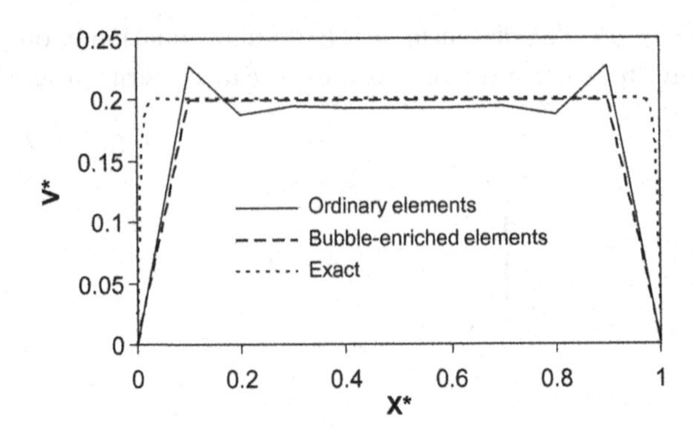

Fig. 5.2 Comparison of velocity profile at $y^* = 7$ generated by the continuous penalty scheme based on ordinary and bubble-enriched elements and analytical solution ($D_a = 4 \times 10^{-4}$ and $v_0^* = 0.2$).

5.3 Solution of the Energy Equation

Using the dimensionless parameters defined by Eq. (5.17) together with the following definition

$$T = T^* T_0, \tag{5.48}$$

where T_0 is a reference value for temperature, the dimensionless form of the two-dimensional steady state energy equation is written as

$$\left(p_{ex} \frac{\partial T^*}{\partial x^*} + p_{ey} \frac{\partial T^*}{\partial y^*} \right) - \left(\frac{\partial^2 T^*}{\partial x^{*2}} + \frac{\partial^2 T^*}{\partial y^{*2}} \right) = 0, \tag{5.49}$$

where P_e is Peclet number defined as

$$P_e = \frac{\rho c_p \mathbf{v} h}{k} \tag{5.50}$$

in which h is a characteristic length, p_{ex} and p_{ey} correspond to x and y components of the velocity vector:

$$P_{ex} = \frac{\rho c_p v_x h}{k}, \quad P_{ey} = \frac{\rho c_p v_y h}{k}.$$

Dimensionless boundary conditions in the rectangular domain shown in Fig. 5.1 are

$$\begin{cases} T^* = 1 \quad \text{for } y^* = 0,\ 0 \le x^* < 1 \text{ and } x^* = 0,\ 0 \le y^* < 10 \\ T^* = 0 \quad \text{for } x^* = 1,\ 0 \le y^* < 10 \\ \dfrac{\partial T^*}{\partial y^*} = 0, \quad \dfrac{\partial T^*}{\partial x^*} = 0 \quad \text{for } y^* = 10,\ 0 \le x^* \le 1 \end{cases} . \tag{5.51}$$

5.3.1 *Finite element scheme used to solve the energy equation*

Following the normal steps of finite element, the field unknown in Eq. (5.49) is approximately represented in terms of a set of shape functions corresponding to the selected element as

$$T^* \approx \tilde{T}^* = \sum_{j=1}^{n} N_j T_j^*, \tag{5.52}$$

where n is the total number of nodes in an element and T_j^* is the nodal value of unknown at the nodes (i.e. sampling points) of an element. After the substitution from Eq. (5.52) into Eq. (5.49) and weighting of the appeared residual the following statement is constructed

$$\int_{\Omega_e} \left[W_i \left(p_{ex} \frac{\partial \sum_{j=1}^{n} N_j T_j^*}{\partial x^*} + p_{ey} \frac{\partial \sum_{j=1}^{n} N_j T_j^*}{\partial y^*} \right) - W_i \left(\frac{\partial^2 \sum_{j=1}^{n} N_j T_j^*}{\partial x^{*2}} + \frac{\partial^2 \sum_{j=1}^{n} N_j T_j^*}{\partial y^{*2}} \right) \right] dx^* dy^* = 0. \tag{5.53}$$

The second-order differentials in Eq. (5.53) are reduced by the application of Green's theorem. This leads to the appearance of boundary integral terms along the exterior boundaries of finite elements. Therefore, corresponding to a total of n interpolation functions, n equations are generated and a system of $n \times n$ equations is constructed. Using matrix notation this system is written as

$$[A_{ij}]\{T_j^*\} = \{B_j\}, \tag{5.54}$$

where

$$A_{ij} = \int_{\Omega_e} \left[W_i \left(p_{ex} \frac{\partial N_j}{\partial x^*} + p_{ey} \frac{\partial N_j}{\partial y^*} \right) + \left(\frac{\partial W_i}{\partial x^*} \frac{\partial N_j}{\partial x^*} + \frac{\partial W_i}{\partial y^*} \frac{\partial N_j}{\partial y^*} \right) \right] dx^* dy^* \tag{5.55}$$

and

$$B_j = \int_{\Gamma_e} W_i \left(\frac{\partial T^{*^e}}{\partial x^*} n_x + \frac{\partial T^{*^e}}{\partial y^*} n_y \right) d\Gamma_e. \tag{5.56}$$

Equation (5.54) provides the working equation of the finite element scheme used to solve the two-dimensional energy equation. In order to simplify repeated application of this equation over the elements of a computational mesh, it is transformed to a local natural coordinate system using isoparametric mapping. Elemental equations are therefore derived and assembled over their common nodes. After the imposition of boundary conditions the derived global set of equations becomes determinate and can be solved.

5.3.2 *Bubble-enriched shape functions used in the solution of the energy equation*

Similar to previously described cases, a set of four shape functions corresponding to a four-node rectangular element is considered here. These shape functions consist of ordinary bilinear shape functions which are enriched by appropriate bubble functions obtained in the previously described manner (Chap. 4) and are expressed as

$$\begin{cases} N_1 = \dfrac{1}{4}(1-\xi)(1-\eta) + b(1-\xi^2)(1-\eta^2) \\ N_2 = \dfrac{1}{4}(1+\xi)(1-\eta) + b(1-\xi^2)(1-\eta^2) \\ N_3 = \dfrac{1}{4}(1+\xi)(1+\eta) + b(1-\xi^2)(1-\eta^2) \\ N_4 = \dfrac{1}{4}(1-\xi)(1+\eta) + b(1-\xi^2)(1-\eta^2) \end{cases}, \tag{5.57}$$

where $b = \frac{l(P_{ex}+P_{ey})}{8}$ (see Eq. (3.114)),
where l is a characteristic element length. It should be emphasized again that in the multiscale finite element scheme only shape functions are enriched by bubble functions and the weight functions remain identical to the ordinary shape functions associated with the selected elements.

As the comparison of equation sets of (5.42) and (5.57) reveals, the bubble-enriched shape functions used to solve flow and energy equations are different. This reflects the underlying difference between the levels of multiscale behavior of these sets of governing equations.

5.3.3 *Solved example — Non-isothermal unidirectional flow*

Consider a porous rectangular domain such as shown in Fig. 5.3. A value of the Darcy number equal to $D_a = 4 \times 10^{-4}$ and a plug flow inlet velocity

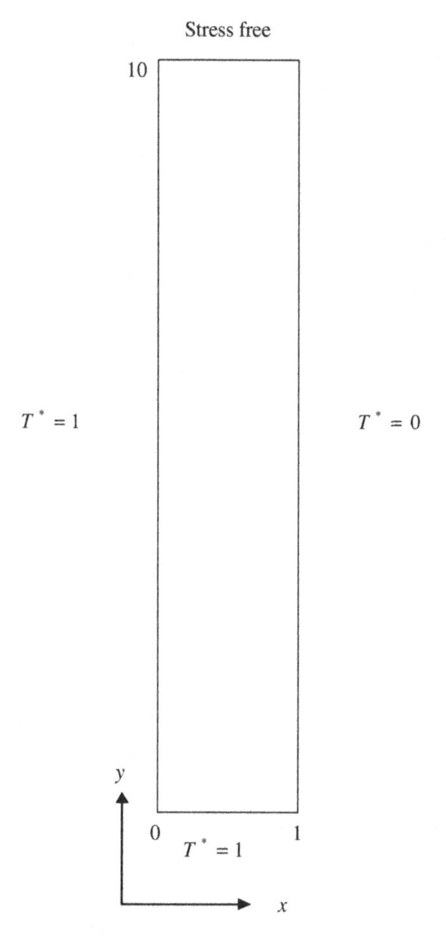

Fig. 5.3 Rectangular domain used to simulate nonisothermal porous flow (temperature boundary conditions are shown, velocity boundary conditions are similar to those shown in Fig. 5.1).

of $v_0^* = 0.2$ are assumed. To assign appropriate values for the physical properties in a porous medium, average values of fluid and solid matrix need to be used (Nield and Bejan, 1992). Here we have chosen the physical values such that the flow Peclet number is

$$\frac{\rho c_p h}{k} = 644.62 \quad \text{and therefore } P_{ex} = 644.62 v_x, \;\; P_{ey} = 644.62 v_y.$$

The value of characteristic h length is chosen to be equal to the domain width and hence $h = 1$. Using these values the bubble function defined as

$\phi_b = b(1-\xi^2)(1-\eta^2)$ becomes

$$\phi_b = \frac{l}{8} 644.62(v_x + v_y)(1-\xi^2)(1-\eta^2), \tag{5.58}$$

where l is a characteristic element length. In Fig. 5.4 the velocity profile at a sample cross-section within the domain is shown. Corresponding temperature profiles obtained using ordinary and bubble function-enriched elements based on the scheme described in Secs. 5.3.1 and 5.3.2 are shown in Fig. 5.5. As shown in Fig. 5.5, the solution obtained using ordinary elements is distorted and unacceptable. In contrast the temperature profile generated by the bubble function-enriched elements is smooth and is as would theoretically be expected for the uniform flow simulated in this example.

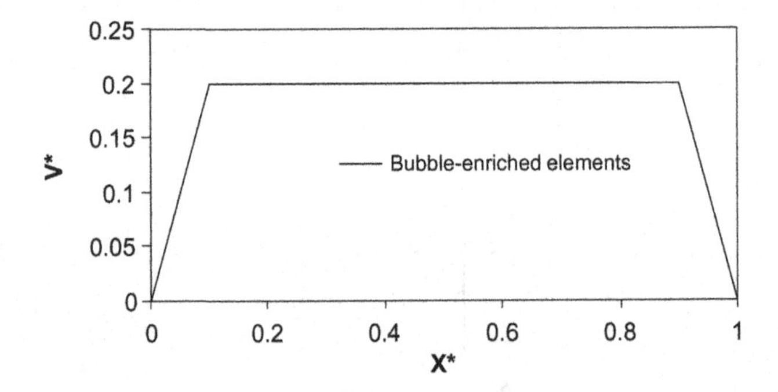

Fig. 5.4 Predicted velocity profile for a nonisothermal flow ($D_a = 4 \times 10^{-4}$, $y^* = 7$).

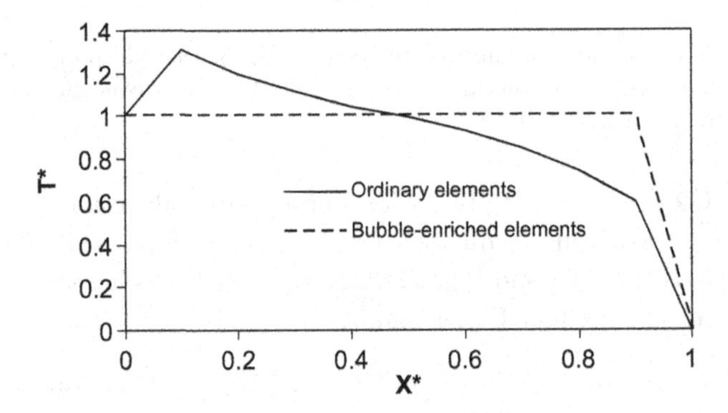

Fig. 5.5 Predicted temperature profile ($D_a = 4 \times 10^{-4}, y^* = 7$).

5.3.4 *Higher-order elements and mesh refinement*

For further evaluation of the performance of bubble function-enriched elements we consider the solution of the described nonisothermal porous flow problem using biquadratic elements, which offer a higher order of approximation than the initially used bilinear elements. We also consider the effects of mesh refinement on the obtained results. The characteristic element length is hence reduced to $l = 0.05$. Figure 5.6 shows the predicted temperature and velocity profiles. Despite the reduction of the inlet velocity from 0.2 (used in the last example) to 0.07 to reduce the importance of convection transport mechanism, an inaccurate temperature profile is predicted. To highlight the importance of this point a repeat solution using the original inlet velocity of 0.2 is also shown in Fig. 5.7. These solutions are repeated using a more refined mesh in which the characteristic element length is reduced to 0.033 (i.e. a 30×30 mesh is used). However, as shown in Fig. 5.8 the predicted temperature profile remains inaccurate and an unstable characteristic of the scheme's inability to cope with "convection dominated" multiscale behavior. To confirm the theoretical consistency of the generated results, a further solution is carried out in which perfect slip wall boundary conditions are imposed (Nassehi, 2002). In this case, because cross-sectional velocity gradient is removed, velocity field remains uniform and as a result temperature field should have significantly greater stability. As shown in Fig. 5.9, in this case the theoretically expected smooth temperature profile is generated.

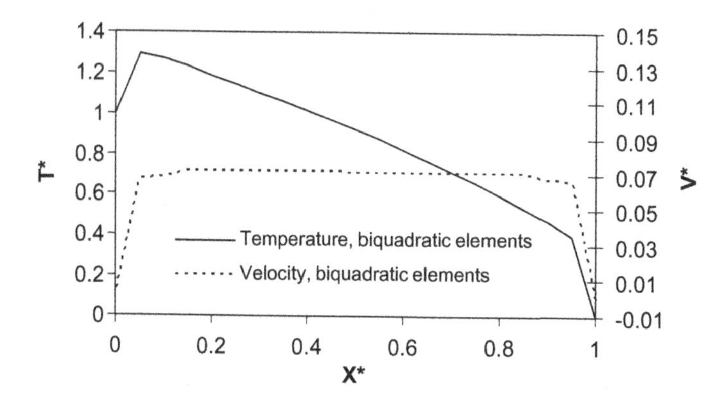

Fig. 5.6 Predicted velocity and temperature profiles (biquadratic elements with $l = 0.05$ inlet velocity $v^* = 0.07, D_a = 4 \times 10^{-4}, y^* = 7$).

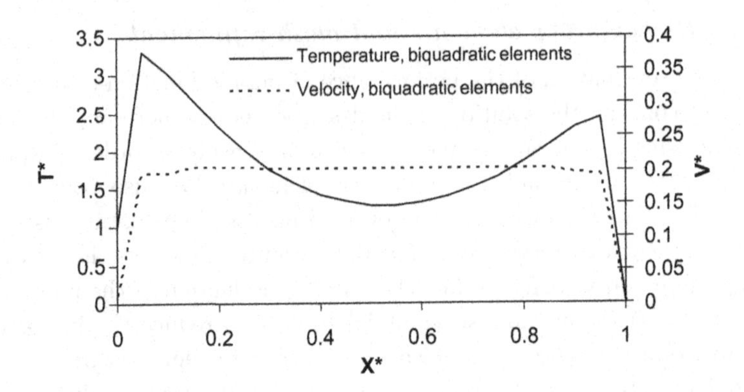

Fig. 5.7 Predicted velocity and temperature profiles (biquadratic elements with $l = 0.05$, inlet velocity $v^* = 0.2, D_a = 4 \times 10^{-4}, y^* = 7$).

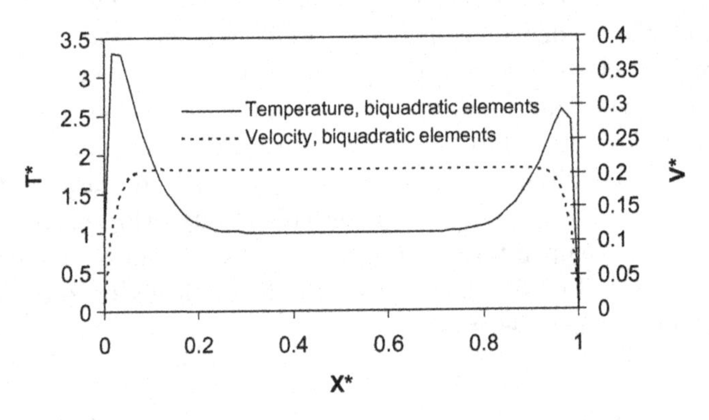

Fig. 5.8 Predicted velocity and temperature profiles (biquadratic elements with $l = 0.033$, inlet velocity $v^* = 0.2, D_a = 4 \times 10^{-4}$, $y^* = 7$).

5.4 Multidirectional, Nonisothermal Porous Flow

We now consider a case in which by imposing a nonuniform inlet velocity, the unidirectional nature of the flow field described in the previous examples is disturbed. However, the rest of the boundary conditions have been kept identical to the previous cases and new results based on the same schemes have been obtained. In this case the inlet condition is defined as

$$\begin{cases} v^* = v_0 & \text{at } 0.1 \le x^* \le 0.9 \\ v^* = 0 & \text{at } 0.0 \le x^* < 0.1. \\ v^* = 0 & \text{at } 0.9 < x^* \le 1.0 \end{cases}$$

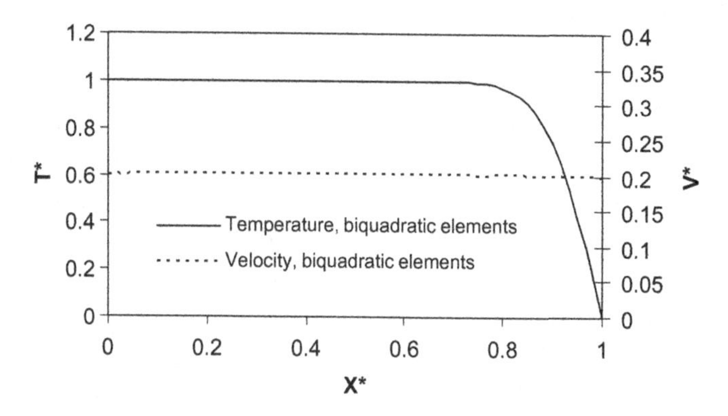

Fig. 5.9 Predicted velocity and temperature profiles (biquadratic elements with $l = 0.033$, inlet velocity $v^* = 0.2$, perfect slip walls).

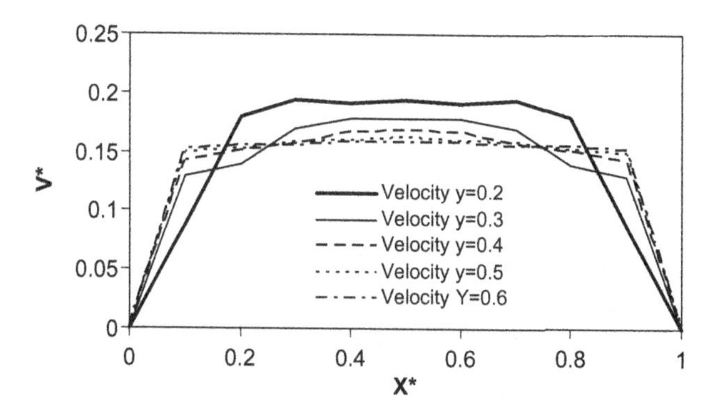

Fig. 5.10 Predictive velocity profiles representing progressive formation of fully developed flow bubble function-enriched elements, $D_a = 4 \times 10^{-4}$, nonuniform, nonisothermal flow.

Results obtained using ordinary and bubble function-enriched bilinear elements are shown in Figs. 5.10–5.12. As shown in these figures, the bubble function-enriched elements always generate theoretically expected results, whilst ordinary bilinear elements fail to generate acceptable solutions.

5.5 Inclusion of Inertia Effects in Porous Flow Models

The simplest form of porous flow equations, i.e. the previously described Darcy's laws breaks down at high velocities. As an alternative to the Darcy equation the following Dupruit (or Forchheimer) equation (Nield and Bejan,

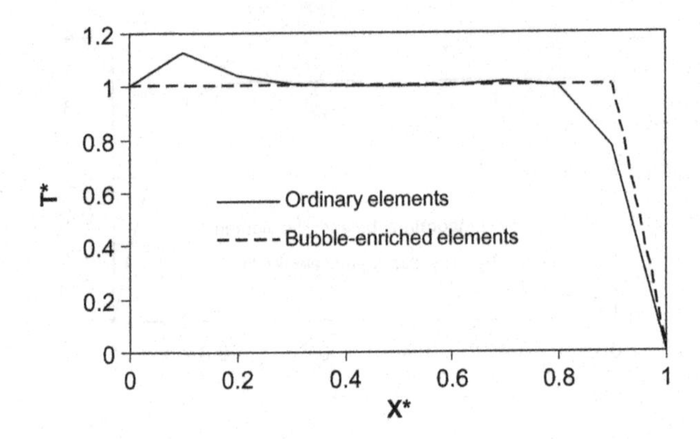

Fig. 5.11 Predicted temperature profiles at $y^* = 1, D_a = 4 \times 10^{-4}$.

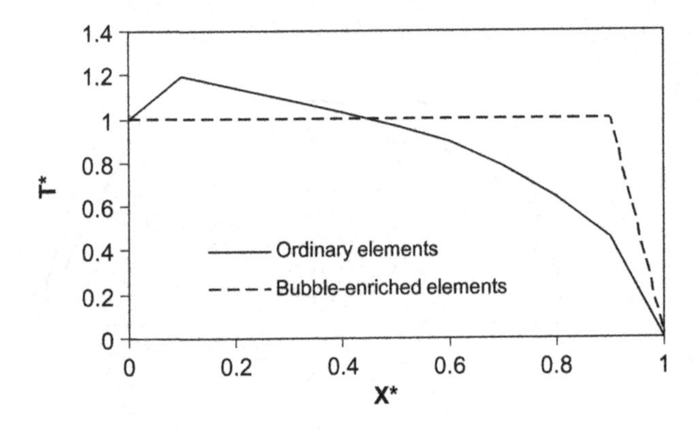

Fig. 5.12 Predicted temperature profile at $y^* = 7, D_a = 4 \times 10^{-4}$.

1992) has been proposed:

$$0 = -\frac{\Delta p}{\Delta x} - \alpha v_x - \beta v_x^2, \tag{5.59}$$

where α and β are parameters which represent the influence of the physical properties of fluid and solid matrix on the flow. The second-order term in Eq. (5.59) represents the form drag imposed on a fluid by a solid surface obstructing its flow path. Using this concept, Eq. (5.59) is rewritten as

$$0 = -\frac{\Delta p}{\Delta x} - \frac{\mu}{K} v_x - C\rho v_x^2, \tag{5.60}$$

where μ is the fluid viscosity, ρ is fluid density, and C is a form coefficient related to geometry of the solid permeable medium with dimension of $\{\text{Length}^{-1}\}$. A fundamental feature of the previously described Brinkman equation is the recognition of a fluid's capability to transmit force by viscous shear independent of viscous drag. By inclusion of viscous shear force in Eq. (5.60), the Brinkman–Hazen–Dupuit–Darcy equation is therefore obtained (Nield and Bejan, 1992) as

$$0 = -\frac{\Delta p}{\Delta x} + \mu \nabla^2 v_x - \frac{\mu}{K} v_x - C\rho v_x^2. \tag{5.61}$$

Using previously defined dimensionless parameters, dimensionless components of Eq. (5.61) in a two-dimensional domain are written as

$$-\frac{\partial p^*}{\partial x^*} - \frac{1}{D_a} v_x^* - \alpha v_x^{*2} + \left(\frac{\partial^2 v_x^*}{\partial x^{*2}} + \frac{\partial^2 v_x^*}{\partial y^{*2}} \right) = 0, \tag{5.62}$$

$$-\frac{\partial p^*}{\partial y^*} - \frac{1}{D_a} v_y^* - \alpha v_y^{*2} + \left(\frac{\partial^2 v_y^*}{\partial x^{*2}} + \frac{\partial^2 v_y^*}{\partial y^{*2}} \right) = 0, \tag{5.63}$$

where D_a is the Darcy parameter defined as $D_a = \frac{K\Phi}{h^2}$ (Φ is porosity of the porous media) and α is a dimensionless shape factor defined as $\alpha = \frac{C\rho^2 g h^4 \Phi}{\mu^2}$.

Multiscale behavior of a porous flow regime which is governed by Eq. (5.61) should by definition be stronger than a Brinkman regime. Therefore, in this section we consider the derivation of a bubble function-enriched scheme for the solution of this equation.

We choose a benchmark rectangular problem domain in which $h_y = 10h_x$ (i.e. a domain identical to the one shown in Fig. 5.1). The flow regime within this domain is assumed to be subject to the following dimensionless boundary conditions:

(I) Entrance

$$v_x^* = 0, \quad v_y^* = v_0^* \quad \text{for } y^* = 0 \text{ and } 0 < x^* < 1, \tag{5.64}$$

here v_0^* is selected to be equal to 0.2.

(II) Impermeable walls

$$\begin{aligned} v_x^* = v_y^* = 0 \quad &\text{for } x^* = 0 \text{ and } 0 \le y^* < 10, \\ v_x^* = v_y^* = 0 \quad &\text{for } x^* = 1 \text{ and } 0 \le y^* < 10. \end{aligned} \tag{5.65}$$

(III) Exit

Stress-free conditions expressed in the dimensionless form are imposed.

$$\tau_{yy}^* = -2\frac{\partial v_y^*}{\partial y^*} = 0 \quad \text{for } y^* = 10 \text{ and } 0 \leq x^* \leq 1, \tag{5.66}$$

$$\tau_{yx}^* = \tau_{xy}^* = -\left(\frac{\partial v_x^*}{\partial y^*} + \frac{\partial v_y^*}{\partial x^*}\right) = 0$$

$$\text{for } y^* = 10 \text{ and } 0 \leq x^* \leq 1, \tag{5.67}$$

$$\tau_{xx}^* = -\frac{\partial v_x^*}{\partial x^*} = 0 \quad \text{for } y^* = 10 \text{ and } 0 \leq x^* \leq 1. \tag{5.68}$$

5.5.1 *Derivation of bubble functions for Brinkman–Hazen–Dupuit–Darcy equation*

The existence of a nonlinear term (i.e. v^2) in Eq. (5.61) results in the appearance of nonlinear nodal unknowns (e.g. in the context of a two-node linear element we need to deal with terms such as $u_1^2, u_2^2, u_3^2, u_1u_2, u_2u_3, u_1u_3$). This precludes the possibility of using the STC procedure explained in Chap. 3 to obtain required bubble coefficients. To resolve this problem we write v^2 as

$$v^2 = Vv, \tag{5.69}$$

where V in a numerical solution procedure is taken to be a known constant parameter (e.g. its value at the end of last iteration cycle). Therefore, to calculate the bubble function coefficient, for example, the x component of Eq. (5.61) can be written as

$$-\frac{\partial p^*}{\partial x^*} - \frac{1}{D_a}v_x^* - \alpha v_x^* V + \left(\frac{\partial^2 v_x^*}{\partial x^{*2}} + \frac{\partial^2 v_x^*}{\partial y^{*2}}\right) = 0. \tag{5.70}$$

Therefore, following the procedure explained in Chap. 3, bubble coefficient for the solution of Eq. (5.61) for a second-order bubble function $(1 - \xi^2)$ is derived as

$$b = \frac{1}{8}\left(1 + \frac{6}{l^2\left(\frac{1}{D_a} + \alpha V\right)}\right)^{-1}. \tag{5.71}$$

5.5.2 *Solved example — High velocity unidirectional porous flow*

After the division of the rectangular domain into a mesh of 100×100 elements, an inlet velocity of $v_y^* = v_0 = 50$ and a Darcy number of $D_a =$

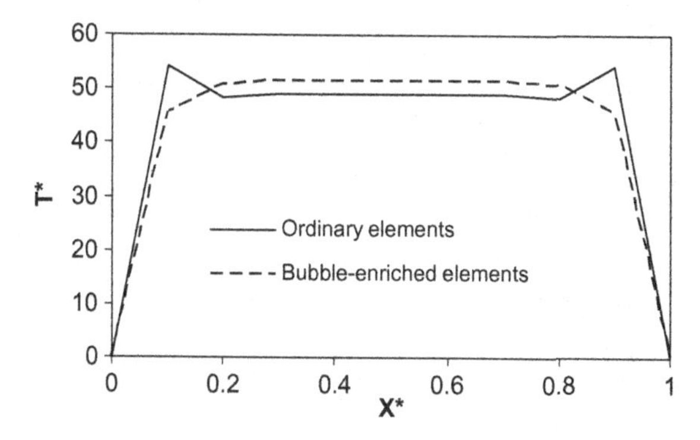

Fig. 5.13 Comparison of predicted velocity profiles ($D_a = 10^{-4}, \alpha = 1000, v_0^* = 50$, $y^* = 7$).

10^{-4} and $\alpha = 1000$ are selected (the remaining boundary conditions are as shown in Eqs. (5.65)–(5.68)). Two sets of results based on schemes which use ordinary bilinear and bubble function-enriched elements are obtained. Comparison of the velocity profiles obtained by these schemes at a cross-section corresponding to $y^* = 7$ is shown in Fig. 5.13. As shown in this figure, the scheme based on the ordinary elements fails to generate accurate results.

5.5.3 *Solved example — Multidirectional, high-velocity porous flow*

We repeat the previous example, this time using the following nonuniform inlet velocity condition (the rest of the parameters and boundary conditions are identical with the case used for unidirectional flow).

$$\begin{cases} v_y^* = v_0^* & \text{at } 0.1 \leq x^* \leq 0.9, \quad v_0^* = 50 \\ v_y^* = 0 & \text{at } 0.0 \leq x^* < 0.1 \text{ and } 0.9 < x^* \leq 1.0 \end{cases} . \tag{5.72}$$

In Fig. 5.14, progressive development of velocity profiles along the rectangular domain at different cross-sections has been shown. These predicted velocity profiles are as would theoretically be expected and confirm the accuracy and consistency of the derived multiscale finite element scheme based on bubble function-enriched elements.

5.6 Solution of Axisymmetric Brinkman Equation

As mentioned earlier, solution of an axisymmetric set of equations provides a three-dimensional simulation for a given problem. Therefore, in cases

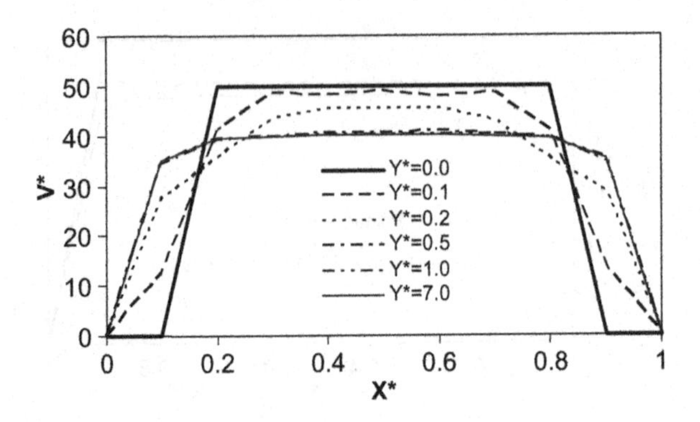

Fig. 5.14 Predictive velocity profiles. ($D_a = 10^{-4}, \alpha = 1000, v_0^* = 50$).

where such a situation can be assumed, solution of axisymmetric equations provides a more realistic result than a two-dimensional approach. Therefore, in this section we consider derivation of multiscale finite element model for axisymmetric Brinkman equation. Using nondimensional forms the governing flow equations in this case are written as

(a) continuity

$$\frac{1}{r^*}\frac{\partial}{\partial r^*}(r^* v_r^*) + \frac{\partial v_z^*}{\partial z^*} = 0, \tag{5.73}$$

(b) r and z components of axisymmetric Brinkman equation

$$-\frac{\partial p^*}{\partial r^*} - \frac{1}{D_a} v_r^* + \left(\frac{\partial}{\partial r^*}\left(\frac{1}{r^*}\frac{\partial}{\partial r^*}(r^* v_r^*)\right) + \frac{\partial^2 v_r^*}{\partial z^{*2}}\right) = 0, \tag{5.74}$$

$$-\frac{\partial p^*}{\partial z^*} - \frac{1}{D_a} v_z^* + \left(\frac{1}{r^*}\frac{\partial}{\partial r^*}\left(r^*\frac{\partial v_z^*}{\partial r^*}\right) + \frac{\partial^2 v_z^*}{\partial z^{*2}}\right) = 0. \tag{5.75}$$

The above set of equations are solved subject to the following dimensionless boundary conditions (see Fig. 5.15)

(I) Inlet

$$v_r^* = 0, \quad v_z^* = v_0^* \quad \text{for } z^* = 0 \text{ and } 0 < r^* < 1, \tag{5.76}$$

here v_0^* is selected to be equal to 0.2.

(II) Impermeable walls

$$\begin{aligned} &\frac{\partial v_z^*}{\partial r^*} = 0, \quad v_r^* = 0 \quad \text{for } r^* = 0 \text{ and } 0 \le z^* < 10, \\ &v_r^* = v_z^* = 0 \quad \text{for } r^* = +0.5 \text{ and } 0 \le z^* < 10. \end{aligned} \tag{5.77}$$

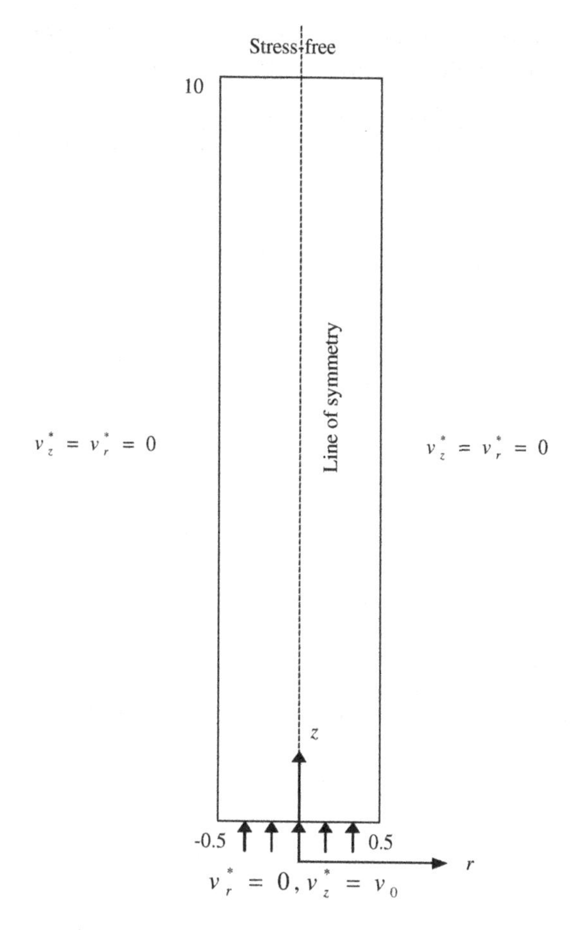

Fig. 5.15 Axisymmetric flow domain and boundary conditions.

(III) Exit

Stress-free conditions expressed in the dimensionless form are imposed.

$$\begin{aligned}
\tau_{zz}^* &= -2\frac{\partial v_z^*}{\partial z^*} = 0 \quad \text{for } z^* = 10 \text{ and } -0.5 \le r^* \le 0.5, \\
\tau_{zr}^* = \tau_{rz}^* &= -\left(\frac{\partial v_r^*}{\partial z^*} + \frac{\partial v_z^*}{\partial r^*}\right) = 0 \quad \text{for } z^* = 10 \text{ and } -0.5 \le r^* \le 0.5, \\
\tau_{rr}^* &= -2\frac{\partial v_r^*}{\partial r^*} = 0 \quad \text{for } z^* = 10 \text{ and } -0.5 \le r^* \le 0.5.
\end{aligned} \tag{5.78}$$

5.6.1 Derivation of the continuous penalty finite element scheme for axisymmetric Brinkman equation

Using a procedure similar to the one explained in Sec. 5.2.1, pressure in the radial and axial components of the axisymmetric Brinkman equation is substituted by

$$p^* = -\lambda_0 \left(\frac{1}{r^*} \frac{\partial}{\partial r^*} \left(r^* v_r^* \right) + \frac{\partial v_z^*}{\partial r^*} \right). \tag{5.79}$$

Therefore

$$\begin{aligned} &-\frac{\partial}{\partial r^*} \left(-\lambda_0 \left(\frac{1}{r^*} \frac{\partial}{\partial r^*} \left(r^* v_r^* \right) + \frac{\partial v_z^*}{\partial r^*} \right) \right) - \frac{1}{D_a} v_r^* \\ &+ \left(\frac{\partial}{\partial r^*} \left(\frac{1}{r^*} \frac{\partial}{\partial r^*} \left(r^* v_r^* \right) \right) + \frac{\partial^2 v_r^*}{\partial z^{*2}} \right) = 0, \end{aligned} \tag{5.80}$$

$$\begin{aligned} &-\frac{\partial}{\partial z^*} \left(-\lambda_0 \left(\frac{1}{r^*} \frac{\partial}{\partial r^*} \left(r^* v_r^* \right) + \frac{\partial v_z^*}{\partial r^*} \right) \right) - \frac{1}{D_a} v_z^* \\ &+ \left(\frac{1}{r^*} \frac{\partial}{\partial r^*} \left(r^* \frac{\partial v_z^*}{\partial r^*} \right) + \frac{\partial^2 v_z^*}{\partial z^{*2}} \right) = 0. \end{aligned} \tag{5.81}$$

Following the procedure described in Sect. 5.2.1 we obtain the following elemental stiffness equation (i.e., the working equation of the scheme)

$$\begin{bmatrix} A_{ij}^{11} & A_{ij}^{12} \\ A_{ij}^{21} & A_{ij}^{22} \end{bmatrix} \begin{Bmatrix} v_{rj}^* \\ v_{zj}^* \end{Bmatrix} = \begin{Bmatrix} B_j^1 \\ B_j^2, \end{Bmatrix}, \tag{5.82}$$

where

$$\begin{aligned} A_{ij}^{11} = \int_{\Omega_e} \Bigg[&\lambda_0 \left(\frac{\partial W_i}{\partial r^*} \frac{\partial N_j}{\partial r^*} + \frac{W_i}{r^*} \frac{N_j}{r^*} + \frac{W_i}{r^*} \frac{\partial N_j}{\partial r^*} + \frac{N_j}{r^*} \frac{\partial W_i}{\partial r^*} \right) \\ &+ \frac{\partial W_i}{\partial r^*} \frac{\partial N_j}{\partial r^*} + \frac{W_i}{r^*} \frac{N_j}{r^*} + \frac{\partial W_i}{\partial z^*} \frac{\partial N_j}{\partial z^*} + \frac{1}{D} W_i N_j \Bigg] r^* dr^* dz^*, \end{aligned}$$

$$A_{ij}^{12} = \int_{\Omega_e} \lambda_0 \left(\frac{\partial W_i}{\partial r^*} \frac{\partial N_j}{\partial z^*} + \frac{W_i}{r^*} \frac{\partial N_j}{\partial r^*} \right) r^* dr^* dz^*,$$

$$A_{ij}^{21} = \int_{\Omega_e} \lambda_0 \left(\frac{\partial W_i}{\partial z^*} \frac{\partial N_j}{\partial r^*} + \frac{N_j}{r^*} \frac{\partial W_i}{\partial z^*} \right) r^* dr^* dz^*,$$

$$A_{ij}^{22} = \int_{\Omega_e} \left[\lambda_0 \left(\frac{\partial W_i}{\partial z^*} \frac{\partial N_j}{\partial z^*} \right) + \frac{\partial W_i}{\partial r^*} \frac{\partial N_j}{\partial r^*} + \frac{\partial W_i}{\partial z^*} \frac{\partial N_j}{\partial z^*} + \frac{1}{D_a} W_i N_j \right] dx^* dy^*,$$

$$B_j^1 = \int_{\Gamma_e} W_i \left(\lambda_0 \left(\frac{1}{r^*} \frac{\partial}{\partial r^*} (r^* v_r^{e*}) + \frac{\partial v_z^{e*}}{\partial r^*} \right) n_r + \left(\frac{\partial v_r^{e*}}{\partial z^*} + \frac{\partial v_r^{e*}}{\partial r^*} \right) n_z \right) r^* d\Gamma_e,$$

$$B_j^1 = \int_{\Gamma_e} W_i \left(\lambda_0 \left(\frac{1}{r^*} \frac{\partial}{\partial r^*} (r^* v_r^{e*}) + \frac{\partial v_z^{e*}}{\partial r^*} \right) n_r + \left(\frac{\partial v_z^{e*}}{\partial z^*} + \frac{\partial v_z^{e*}}{\partial r^*} \right) n_z \right) r^* d\Gamma_e.$$

The second-order differentials in the weighted residual statement are again reduced using Green's theorem (in its general form called the divergence theorem, see Appendix A.4) written as

$$\int_\Omega w \vec{\nabla}.\vec{A} d\Omega = \oint_\Gamma w \vec{A}.\vec{n} d\Gamma - \int_\Omega \nabla \vec{w} \cdot \vec{n} d\Omega. \tag{5.83}$$

In a cylindrical system we have

$$\vec{\nabla} \cdot \vec{A} = \frac{1}{r} \frac{\partial}{\partial r} (r A_r) + \frac{\partial A_z}{\partial z}. \tag{5.84}$$

To maintain consistency with Eq. (5.84), the radial derivative of pressure is written as

$$\frac{\partial p}{\partial r} = \frac{1}{r} \frac{\partial}{\partial r} (rp) - \frac{p}{r}. \tag{5.85}$$

5.6.2 *Derivation of bubble functions for axisymmetric Brinkman equation*

Consider the following set of equations which represent one-dimensional Brinkman flow in r and z directions, respectively.

$$\text{In } r^* \text{direction: } -\frac{\partial p^*}{\partial r^*} - \frac{1}{Da} v_r^* + \frac{\partial}{\partial r^*} \left(\frac{1}{r^*} \frac{\partial}{\partial r^*} (r^* v_r^*) \right) = 0. \tag{5.86}$$

$$\text{In } z^* \text{ direction: } -\frac{\partial p^*}{\partial z^*} - \frac{1}{D_a} v_z^* + \frac{\partial^2 v_z^*}{\partial z^{*2}} = 0. \tag{5.87}$$

The z^* component is identical to a Cartesian form and hence previously described techniques to derive bubble functions for both RFB and STC techniques can be used (Chap. 3). In r^* direction, however, a different situation is encountered. This equation can be written as

$$-\frac{\partial p^*}{\partial r^*} - \frac{1}{D_a} v_r^* + \frac{1}{r^*} \frac{\partial v_r^*}{\partial r^*} - \frac{v_r^*}{r^{*2}} + \frac{\partial^2 v_r^*}{\partial r^{*2}} = 0. \tag{5.88}$$

We now consider the weighted residual statement based on a differential operator in a cylindrical coordinate system:

$$\int_{\Omega_e} w L(u) d\Omega_e = \int_{\Omega_e} w L(u) r d\theta dr dz. \tag{5.89}$$

The reduction to a one-dimensional form in the radial direction requires that

$$\int_0^{2\pi}\int_0^l\int_0^l wL(u)rd\theta drdz = 2\pi\int_0^l wL(u)rdr. \tag{5.90}$$

Therefore, the weighted residual statement corresponding to Eq. (5.88) is given as

$$\int_0^l w\left(-\frac{1}{Da}v_r^* + \frac{1}{r^*}\frac{\partial v_r^*}{\partial r^*} - \frac{v_r^*}{r^{*2}} + \frac{\partial^2 v_r^*}{\partial r^{*2}}\right)r^*dr^* = 0. \tag{5.91}$$

Using a local natural coordinate system of $\xi(+1,-1)$, however, we have

$$r^* = \frac{l}{2}(1+\xi) \tag{5.92}$$

and

$$\int_{-1}^{+1} w\left(-\frac{l^2}{4}(1+\xi)\frac{1}{D_a}v_r^* + \frac{\partial v_r^*}{\partial \xi} - \frac{1}{(1+\xi)}v_r^* + (1+\xi)\frac{\partial^2 v_r^*}{\partial \xi^2}\right)d\xi = 0. \tag{5.93}$$

Using Green's theorem and rearranging the above equation

$$\int_{-1}^{+1}\left(-(1+\xi)\frac{\partial w}{\partial \xi}\frac{\partial v_r^*}{\partial \xi} + w\frac{\partial v_r^*}{\partial \xi} - \frac{l^2}{4}(1+\xi)\frac{1}{D_a}wv_r^* - \frac{1}{(1+\xi)}wv_r^*\right)d\xi = 0. \tag{5.94}$$

In a "two node" bubble function-enriched approximation of the dimensionless velocity is written as

$$\begin{aligned} v_r^* &= \psi_1 v_{r1}^* + \psi_2 v_{r2}^* + \phi_b v_{r3}^* \\ &= \frac{1}{2}(1-\xi)v_{r1}^* + \frac{1}{2}(1+\xi)v_{r2}^* + (1-\xi^2)v_{r3}^*. \end{aligned} \tag{5.95}$$

Choosing $w = (1-\xi^2)$

$$\begin{aligned} \int_{-1}^{+1} &\left\{-(1+\xi)(-2\xi)\left(-\frac{1}{2}v_{r1} + \frac{1}{2}v_{r2} - 2\xi v_{r3}\right)\right. \\ &+ (1-\xi^2)\left(-\frac{1}{2}v_{r1} + \frac{1}{2}v_{r2} - 2\xi v_{r3}\right) \\ &- \frac{l^2}{4}\frac{1}{Da}(1+\xi)(1-\xi^2)\left(\frac{1}{2}(1-\xi)v_{r1} + \frac{1}{2}(1+\xi)v_{r2} + (1-\xi^2)v_{r3}\right) \\ &\left. - \frac{(1-\xi^2)}{(1+\xi)}\left(\frac{1}{2}(1-\xi)v_{r1} + \frac{1}{2}(1+\xi)v_{r2} + (1-\xi^2)v_{r3}\right)\right\} d\xi = 0. \end{aligned} \tag{5.96}$$

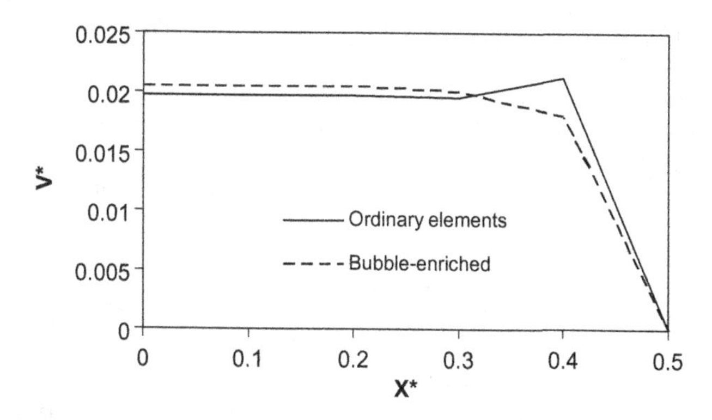

Fig. 5.16 Comparison of velocity profiles generated using ordinary and bubble function-enriched elements (domain and boundary conditions based on Fig. 5.15, $D_a = 10^{-3}$, STC-enriched cylindrical elements, $y^* = 7$).

After integration we have

$$v_{r3} = -\frac{\frac{13}{3} + \frac{2l^2}{15D_a}}{\frac{128}{30} + \frac{4l^2}{15D_a}} v_{r1} + \frac{\frac{2}{3} - \frac{l^2}{5D_a}}{\frac{128}{30} + \frac{4l^2}{15D_a}} v_{r2}. \tag{5.97}$$

Therefore,

$$b_1 = -\frac{\frac{13}{3} + \frac{2l^2}{15D_a}}{\frac{128}{30} + \frac{4l^2}{15D_a}}, \quad b_2 = \frac{\frac{2}{3} - \frac{l^2}{5D_a}}{\frac{128}{30} + \frac{4l^2}{15D_a}}. \tag{5.98}$$

A sample result obtained using the described scheme and its comparison with the solution generated using ordinary elements is shown in Fig. 5.16. As shown in this figure, the scheme using axisymmetric bubble function-enriched elements generates a stable, theoretically expected result whilst the result obtained by the ordinary elements is unacceptable. This solution provides confirmation for the consistency of multiscale approach described in this section for an axisymmetric porous flow problem.

References

[1] Allan, F.M. and Hamdan, M.H., Fluid mechanics of the interface region between two porous layers, *Appl. Math. Comput.*, 2002; 128; 37–43.

[2] Bai, W., The quadrilateral 'Mini' finite element for the stokes problem, *Comput. Meth. Appl. Mech. Eng.*, 1997; 143; 41–47.

[3] Gerald, C.F. and Wheatley, P.O., *Applied Numerical Analysis*, Addison-Wesley, 1984.

[4] Hsu, C.T. and Cheng, P., The Brinkman model for natural convection about a semi-infinite vertical flat plate in a porous medium, *Int. J. Heat Mass Transfer*, 1985; 28; 683–697.

[5] Irons, B.M., A frontal solution program for finite element analysis, *Int. J. Numer. Meth. Eng.*, 1970; 2; 5–32.

[6] Kaviany, M., Non-Darcian effects on natural convection in porous media confined between horizontal cylinders, *Int. J. Heat Mass Transfer*, 1986; 29; 1513–1519.

[7] Nassehi, V., *Practical Aspects of Finite Element Modelling of Polymer Processing*, John Wiley & Sons, Chichester, 2002.

[8] Nield, D.A. and Bejan, A., *Convection in Porous Media*, Springer-Verlag, New York, 1992.

[9] Parvazinia, M., Nassehi, V., Wakeman, R.J. and Ghoreishy, M.H.R., Finite element modelling of flow through a porous medium between two parallel plates using the Brinkman equation, *Transport Porous Media*, 2006a; 63; 71–90.

[10] Parvazinia, M., Nassehi, V. and Wakeman, R.J., Multi–scale finite element modeling using bubble function method for a convection–diffusion problem, *Chem. Eng. Sci.*, 2006b; 61; 2742–2751.

[11] Pittman, J.F.T., Finite elements for field problems, in *Computer Modelling for Polymer Processing*, Tucker III, C.L. (Eds.), Hanser Publishers, Munich, 1989.

[12] Reddy, J.N., *Applied Functional Analysis and Variational Methods in Engineering*, McGraw-Hill, New York, 1986.

[13] Zienkiewicz, O.C. and Taylor, R.L., *The Finite Element Method*, McGraw-Hill, London, 1994.

CHAPTER 6

Computer Program

In this chapter, details of the finite element code used to solve two-dimensional multiscale transport equations presented in this book are described. This code can be used to repeat and verify results of the solved examples and to conduct further numerical experiments. The program starts with comment statements defining the used variables. Further comment statements are included within the program to make its understanding and use easier for readers. The "Code" is written in FORTRAN and its implementation requires a suitable FORTRAN compiler. However, the use of specialized forms is avoided to make it compatible with most types of FORTRAN compilers that are in the public domain. Sample input and output files are included to guide the user in the preparation of new input files and interpretation of the results of computations. Through the comparison of sample "input" and "output" all necessary information about that program can be obtained.

6.1 Program Structure

The program starts by asking the input and output file names. From input file it reads the total number of elements and nodes, the number of nodes per element, Peclet (P_e) and Damköhler (D_a) numbers, boundary conditions, nodal coordinates, and finally, element connectivity data. The main part of the computer code consists of the calculations related to the formulation of stiffness matrix. This task is performed by a subroutine called STIFFT. Shape functions are accessed via subroutine SHAPEB which is called within subroutine STIFFT. A sample of bubble-enriched shape functions is written within the program as working examples. Other types of bubble functions can be derived using the explanations given in previous chapters and included within this subroutine. Calculation of the element characteristic length required in the bubble function coefficient is performed

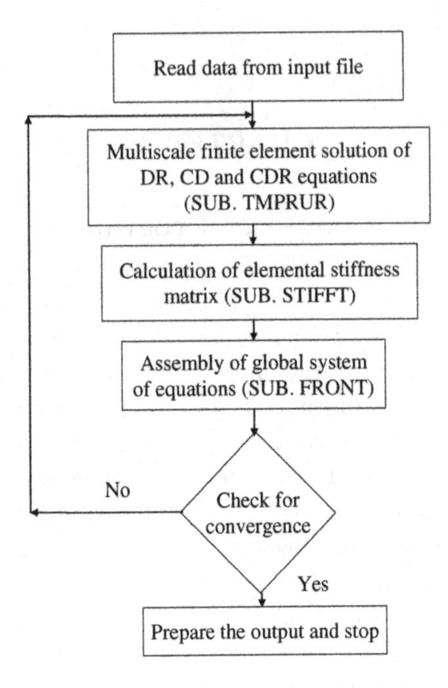

Fig. 6.1 Schematic diagram of the computer code.

within a subroutine called ELMTLNTH. After calculation of elemental stiffness matrix they are assembled into a global matrix, boundary conditions are imposed and numerical solution is performed in subroutine FRONT. A schematic diagram of the structure of the program is shown in Fig. 6.1.

6.2 Source Code

```
C.....................................................................
C   LIST OF SYMBOLS                                                   .
C.....................................................................
C
C   AK. . . . . THERMAL CONDUCTIVITY
C   BC. . . . . PRIMARY B.C. ARRAY
C   BCT . . . . PRIMARY B.C. ARRAY  (TEMPERATURE)
C   CP. . . . . HEAT CAPACITY
C   ELSTIF. . . ELEMENTAL STIFFNESS MATRIX
C   EQ. . . . . A VARIABLE USED IN FRONTAL ROUTINE (GLOBAL MATRIX)
C   GAUSS . . . GAUSS POINTS ARRAY
C   IF. . . . . FULL INTEGRATION INDEX
C   IR. . . . . REDUCED INTEGRATION INDEX
C   JMOD. . . . A VARIABLE USED IN FRONTAL ROUTINE
C   LDEST . . . A VARIABLE USED IN FRONTAL ROUTINE
```

```
C   LHED. . . . A VARIABLE USED IN FRONTAL ROUTINE
C   LPIV. . . . A VARIABLE USED IN FRONTAL ROUTINE
C   MAXFR . . . SIZE OF ARRAYS IN FRONTAL ROUTINE
C   MDF . . . . NO. OF D.O.F. AT EACH NODE
C   MDFT. . . . NO. OF D.O.F. AT EACH NODE (TEMPERAT) , ARRAY
C   MRCL. . . . RECORDED FILE LENGTH (DEFAULT) FOR FRONTAL ROUTINE
C   NBD1. . . . NO. OF FIRST TYPE B.C. NODES
C   NBF . . . . NO. OF THIRD TYPE B.C. ELEMENTS
C   NCOD. . . . CODE FOR PRIMARY B.C.
C   NCODT . . . CODE FOR PRIMARY B.C. (TEMPERAT) , ARRAY
C   NDFT. . . . NO. OF D.O.F. AT EACH NODE (TEMPERAT) (COMPARE WITH
C  MDFT)
C   NDNT. . . . TOTAL NO. OF D.O.F. AT EACH ELEMENT (TEMPERAT)
C   NELM. . . . MAX. NO. OF ELEMENTS
C   NEM . . . . NO. OF ELEMENTS
C   NEQ . . . . NO. OF VELOCITY FIELD EQUATIONS
C   NET . . . . NO. OF TEMPERATURE FIELD EQUATIONS
C   NITER . . . MAX. NO. OF PERMITTED ITERATIONS
C   NNEE. . . . NUMBERS OF ELEMENTS ATTACHED TO EACH NODE, ARRAY
C   NNM . . . . NO. OF NODES
C   NNOD. . . . MAXIMUM NO. OF NODES
C   NOD . . . . ELEMENT CONNECTIVITY MATRIX
C   NOP . . . . ELEMENT CONNECTIVITY MATRIX (FOR PRE-FRONT)
C   NOPP. . . . CODED VALUE OF THE FIRST D.O.F. AT EACH NODE
C   NOPPT . . . CODED VALUE OF THE FIRST D.O.F. AT EACH NODE
C   (TEMPERAT)
C   NPE . . . . NUMBER OF NODES PER ELEMENTS
C   NSIZ. . . . SIZE OF THE GLOBAL VECTOR OF UNKNOWNS
C   NSTF. . . . SIZE OF ELEMENT STIFFNESS MATIX AND LOAD VECTOR
C   NTIME . . . NO. OF TIME STEPS
C   NTRAN . . . =0  STEADY STATE  ,   NE.0  TRANSIENT
C   PVKOL . . . A VARIABLE FOR FRONTAL ROUTINE
C   QQ. . . . . A VARIABLE FOR FRONTAL ROUTINE
C   R1. . . . . A VARIABLE FOR FRONTAL ROUTINE
C   SO. . . . . VELOCITY - PREVIOUS TIME STEP VECTOR
C   TF. . . . . TEMPERATURE - CURRENT SOLUTION VECTOR
C   TFA . . . . TEMPERATURE - INITIAL VECTOR
C   TFI . . . . TEMPERATURE - INITIAL VECTOR (2)
C   TIME. . . . TIME
C   TINIT . . . INITIAL TEMPERATURE
C   TN. . . . . ERROR NORM FOR TEMPERATURE
C   TO. . . . . TEMPERATURE - PREVIOUS TIME STEP VECTOR
C   WT. . . . . WEIGHTS ARRAYS FOR GAUSSIAN QUADRATURE
C   X . . . . . X-COORDINATE (CARTESIAN)
C   Y . . . . . Y-COORDINATE (CARTESIAN)
C
C
C   NOTE : ANY SYMBOL NOT DEFINED IN THE ABOVE LIST IS LOCALLY
C     DEFINED
C.............................................................
```

```
C.........................................................
C  LIST OF SUBROUTINES                                    .
C.........................................................

C ANODAE. . FIND THE CONNECTIVITY OF EACH NODE TO ADJACENT
C ELEMENTS
C ARR2ZF. . INITIALIZE TWO DIMENSIONAL ARRAYS
C ARRZRI. . INITIALIZE ONE DIMENSIONAL ARRAYS (INTEGER)
C BACSUB. . BACK SUBSTITUTION FOR FRONTAL ALGORITHM
C FRONT . . FRONTAL SOLVER
C JACOB2. . CALCULATION OF THE JACOBIAN MATRIX AND ITS
C DETERMINANT
C OUTPUT. . WRITE OUTPUT
C PREFNT. . PRE-FRONT ROUTIE
C RENWAL. . RENEWING OF THE SOLUTION VECTOR USING OVER-
C RELAXATION METHOD
C RSAVE . . LEFT HAND SIDE EQUAL TO THE RIGHT HAND SIDE
C RSAVI . . LEFT HAND SIDE EQUAL TO THE RIGHT HAND SIDE  (INTEGER)
C SHAPE . . SHAPE FUNCTIONS AND THEIR DERIVATIVES
C STIFFT. . STIFFNESS MATRIX CALCULATION  (HEAT)
C TMPRUR. . TEMPERATURE CALCULATION
C VNORM . . CALCULATION OF THE ERROR NORM
C.......................................................................
C  INCLUDE 'SIZEF'
      PARAMETER (NELM  = 40000)
      PARAMETER (NNOD  = 160000)
      PARAMETER (NSTF  = 18)
      PARAMETER (NSIZ  = 400000)
      PARAMETER (MAXFR = 2500)
      PARAMETER (MRCL  = 10024)
C
C....  NELM   MAXIMUM NO. OF ELEMENTS
C....  NNOD   MAXIMUM NO. OF NODES
C....  NSTF  SIZE OF ELEMENT STIFFNESS MATIX AND LOAD VECTOR
C....  NSIZ SIZE OF GLOBAL UNKNOWNS
C....  MAXFR  SIZE OF ARRAYS IN FRONTAL SUB.
C....  MRCL SIZE OF RECORD LENGTH FOR FRONTAL ROUTINE FILE ( TO FIX
C   THE  SCRATCH FILE CORRECTLY
     C IMPLICIT REAL*8 (A-H,O-Z)
C
C.... STRING VARIABLE DECLARATION
C
      CHARACTER TITLE*60
      CHARACTER CT*2
      CHARACTER FNAME1*30, FNAME2*30, FNAME3*30
C
C.... STORAGE ALLOCATION
C
      DIMENSION NOD  (NELM, 9), NOP   (NELM, 9)
      DIMENSION X    (NNOD), Y     (NNOD)
```

```
C.............
      DIMENSION CORP (NNOD, 2), AINIT (NSIZ, 5)
C.............
      DIMENSION TF   (NSIZ), TFI (NSIZ), TFA (NSIZ), THF(NSIZ)
      DIMENSION SO    (NSIZ), TO (NSIZ)
      DIMENSION NOPPV (NSIZ), MDFV  (NSIZ)
      DIMENSION NCODT (NSIZ), BCT   (NSIZ), NOPPT (NSIZ),
     1          MDFT  (NSIZ)
      DIMENSION NCOD  (NSIZ), BC    (NSIZ), NOPP  (NSIZ),
     1          MDF   (NSIZ)
      DIMENSION NDNV  (NSIZ), NDNT  (NSIZ), NNEE (NNOD)
      DIMENSION ELSTIF (NSTF, NSTF), ELF (NSTF), ELF1 (NSTF)

C....           FRONTAL ROUTINE  VARIABLES

      DIMENSION LDEST (NSTF),
     1          LHED  (MAXFR),
     2          NK    (NSTF),
     3          LPIV  (MAXFR),
     4          JMOD  (MAXFR), QQ    (MAXFR),
     5          PVKOL (MAXFR), R1    (NSIZ),
     6          EQ    (MAXFR, MAXFR)
C
C....           GAUSSIAN QUADRATURE INTEGRATION VARIABLES
C
      DIMENSION GAUSS (7, 7), WT (7, 7)

C..........................................................
C     GAUSSIAN QUADRATURE INTEGRATION DATA                .
C..........................................................

      DATA GAUSS/7*0.0D0,.57735027D0,-.57735027D0,5*0.0D0,-.77459667D0,
     *0.0D0,.77459667D0,4*0.0D0,-.8611363116D0,-.3399810435D0,
     *.3399810435D0,.8611363116D0,3*0.0D0,
     *-.90617985D0,-.53846931D0,0.0D0,.53846931D0,.90617985D0,2*0.0D0,
     *-.93246951D0,-.66120939D0,
     *-.23861918,.23861918D0,.66120939D0,.93246951D0,0.0D0,
     *-.94910791D0,-.74153119D0,-.40584515D0,0.0D0,.40584515D0,
     *.74153119D0,.94910791D0/

       DATA WT/2.0D0,6*.0D0,2*1.0,5*.0D0,.55555555D0,.88888888D0,
     *.55555555D0,4*0.0D0,
     *.3478548451D0,.6521451548D0,.6521451548D0,.3478548451D0,3*0.0D0,
     *.23692689D0,.47862867D0,.56888889,.47862867D0,.23692689D0,2*0.0D0,
     *.17132449D0,.36076157D0,.46791393D0,.46791393D0,.36076157D0,
     *.17132449D0,0.0D0,
     *.12948497D0,.27970539D0,.38183005D0,.41795918,.38183005D0,
     *.27970539D0,.12948497D0/

C       OPEN FILES FOR I/O                   .
```

```
C...........................................
      OPEN(UNIT=14,FORM='UNFORMATTED'
                                    ,STATUS='SCRATCH',RECL=MRCL)
      OPEN(UNIT=10 ,FILE='FERROR.LOG', FORM='FORMATTED')
C.................................................
C     READING OF THE INPUT DATA  & PREPROCESSING  .
C.................................................
      WRITE  (*, 5000)
      READ   (*, 6000) FNAME1
C
      WRITE  (*, 5003)
      READ   (*, 6000) FNAME2
C
      OPEN (UNIT=1, FILE=FNAME1, FORM='FORMATTED')
      OPEN (UNIT=2, FILE=FNAME2, FORM='FORMATTED')
C
      PRINT *
      PRINT *,' READING INITIAL AND CONTROL DATA'
      READ   (1, 6010)  TITLE
      WRITE  (2, 5025)  TITLE
C........................
      READ   (1, 6015)  NEM, NNM, NPE, NBD1, NITER
      READ   (1, 6020)   PE, PD
C............................................
C     CHECK THE NO. OF ELEMENTS AND NODES    .
C............................................
      IF (NNM .GT. NNOD) THEN
         WRITE (*, 5045) NNOD
         STOP
      ELSEIF (NEM .GT. NELM) THEN
         WRITE (*, 5050) NELM
         STOP
      ENDIF
C............................................
C     EVALUTIONS OF NDN, NOPP, MDF          .
C............................................
      NDFT = 1
      CALL ARRZRI (NDNT, NSIZ)
      CALL ARRZRI (MDFT, NSIZ)
      CALL ARRZRI (NOPPT, NSIZ)

C...
      DO I= 1, NELM
         NDNT (I) = NPE*NDFT
      ENDDO
      K=-1
      DO I= 1, NNM
         K=K+2
         MDFT  (I) = NDFT
```

```
        NOPPT (I) = I

      ENDDO

C.........................................
C     PRIMARY DEGREE OF FREEDOM          .
C.........................................
      CALL ARRZRI (NCODT , NSIZ)
C....
      IF (NBD1  .NE.  0) THEN
C.....
          DO 10 I= 1, NBD1
             READ   (1, 6025) NODP,CT,VAL3

C...............................
             IF (CT.EQ.'T') THEN
                NCODT (NODP) = 1
                BCT   (NODP) = VAL3
             ENDIF
C...............................
 10       CONTINUE
C.....
      ENDIF
C.........................................
C.........................
      NET = NNM
C............................................

      DO 14 I= 1, NNM
         READ  (1, 6030)     X(I), Y(I)
 14   CONTINUE
C.........................................
C     CONNECTIVITY DATA                  .
C.........................................
      DO 16 I=1, NEM
         READ  (1, 6035) J, (NOD (J,K), K= 1, NPE)
 16   CONTINUE
C.................................................
C     FIND LAST APPEARANCE OD EACH NODE (PRE-FRONT).
C.................................................
      DO I= 1,NEM
         DO J=1,NPE
            NOP (I,J) = NOD (I,J)
         ENDDO
      ENDDO
      CALL   PREFNT  (NNM, NEM, NOP, NPE, NELM)
C...........................................
450   CLOSE (1)
C............................................
```

```
C  DETERMINATION OF INTG. INDEX                .
C...............................................
      IF (NPE.EQ.4) THEN
             IF=7
      ELSEIF (NPE.EQ.8.OR.NPE.EQ.9) THEN
             IF=3
      ENDIF

C.....................................................
      PRINT *,' START OF FINITE ELEMENT CALCULATION   '
C.....................................................
      DO 60 NT= 1,1

      DO 1000 ITER = 1,NITER

C................................................
C     CALCULATION OF TRANSPORT EQUATIONS         .
C................................................
      CALL TMPRUR (
     1  TF, TFI, TO, SO,
     2  GAUSS, WT,
     3  X, Y, NOD, NOP,
     4  BCT, NCODT, NOPPT, MDFT, NDNT,
     5  NPE, IR, IF, DT, NDFT, NEM,
     6  NET, NNM, NTRAN,
     7  PD, PE,
     8  NSIZ, NSTF, NELM, NNOD, MAXFR, IVIS,
     9  R1, ELF, ELSTIF, ELF1,
     1  LDEST, NK, EQ, LHED, LPIV,
     2  JMOD, QQ, PVKOL,
     3  NCOD, BC, NOPP, MDF)
C................................................

      CALL VNORM (TN, NET, TFA, TF, NSIZ)
      WRITE (*, 5115) TN
      WRITE (10, 5115) TN

C..............................................
C     CHECK THE CONVERGENCE                    .
C..............................................

      CALL RENWAL (TFI, TF, NET, 1.00D00)

C...............................................
C     OUTPUTING THE NODAL VALUES                .
C...............................................
      CALL       OUTPUT (NNM, TF, NSIZ)

C...................................................................
 1000 CONTINUE
```

```
C.................................
  60  CONTINUE
C.................................
      CLOSE (2)
      CLOSE (1)
      CLOSE (10)
      CLOSE (14)
C..................................................
 5000 FORMAT (1X,'ENTER INPUT  FILE NAME ___  ',$)
 5003 FORMAT (1X,'ENTER OUTPUT FILE NAME ___  ',$)
 5025 FORMAT (1X,'TITLE : ',A,/)
 5045 FORMAT (1X,/,'--ERROR-- MAXIMUM NO. OF NODES ',I7)
 5050 FORMAT (1X,/,'--ERROR-- MAXIMUM NO. OF ELEM. ',I7)
 5065 FORMAT (1X,/,'--ERROR IN NO. OF INITIAL DEFINITION OF EQUATIONS')
 5115 FORMAT (' >E(TP)=',F7.4,$)
 5120 FORMAT (' >E(FS)=',F7.4,$)
 5125 FORMAT (' >E(CR)=',F7.4,$)
C
C...  READ FORMATS ...................
C
 6000 FORMAT (A)
 6010 FORMAT (A)
 6015 FORMAT (16I5)
 6020 FORMAT (2D15.5)
 6025 FORMAT (I5,6X,A2,F10.6)
 6030 FORMAT (5X,2G20.8)
 6035 FORMAT (10I5)
 6040 FORMAT (3F10.4)
C
      STOP
      END
C..............................................................
C     END OF THE MAIN PROGRAM                                 .
C.............................................
C    SOLUTION OF TRANSPORT EQUATION          .
C.............................................
      SUBROUTINE TMPRUR (
     1  TF, TFI, TO, SO, GAUSS, WT, X, Y, NOD, NOP,
     4  BCT, NCODT, NOPPT, MDFT, NDNT, NPE, IR, IF,
     5  DT, NDFT, NEM, NET, NNM, NTRAN, PD, PE,
     8  NSIZ, NSTF, NELM, NNOD, MAXFR, IVIS, R1, ELF,
     9  ELSTIF, ELF1, LDEST, NK, EQ, LHED, LPIV,
     2  JMOD, QQ, PVKOL, NCOD, BC, NOPP, MDF)
C..............................
      IMPLICIT REAL*8 (A-H,O-Z)
C..............................
      DIMENSION TF    (NSIZ), TFI (NSIZ), TO (NSIZ), SO  (NSIZ)
      DIMENSION GAUSS (7,7), WT(7,7)
      DIMENSION X     (NNOD), Y (NNOD)
      DIMENSION NOD   (NELM, 9), NOP (NELM, 9)
```

```
      DIMENSION BCT   (NSIZ), NCODT (NSIZ), NOPPT (NSIZ)
      DIMENSION MDFT  (NSIZ), NDNT  (NSIZ)
C..............................
      DIMENSION ELF   (NSTF), ELSTIF (NSTF,NSTF)
      DIMENSION DMASS (NSTF, NSTF), ELF1 (NSTF)
      DIMENSION LDEST (NSTF), LHED (MAXFR)
      DIMENSION NK    (NSTF), LPIV (MAXFR)
      DIMENSION JMOD  (MAXFR), QQ (MAXFR)
      DIMENSION PVKOL (MAXFR), R1 (NSIZ)
      DIMENSION EQ    (MAXFR, MAXFR)
      DIMENSION BC    (NSIZ), NCOD (NSIZ), NOPP (NSIZ)
      DIMENSION MDF   (NSIZ)
C............................
      CALL RSAVE (BC, BCT, NSIZ)
      CALL RSAVI (NCOD, NCODT, NSIZ)
      CALL RSAVI (NOPP, NOPPT, NSIZ)
      CALL RSAVI (MDF, MDFT, NSIZ)
C.....
      DO 34 NE=1,NEM
         CALL STIFFT  (NE, NPE, GAUSS, WT, PD, ELSTIF,
     1                PE, NOD, X, Y, IR,
     2                 IF, NELM, NNOD, NSIZ, NSTF)
         CALL FRONT
     1(ELSTIF, ELF, NE, NOP, NELM, NSTF, LDEST, NK,
     2  MAXFR, EQ, LHED, LPIV, JMOD, QQ, PVKOL, TF,
     3  R1, NCOD, BC, NOPP, MDF, NDNT, NSIZ, NEM,
     4  NSIZ, NET, LCOL, NELL, NPE)

   34 CONTINUE
C....................
      RETURN
C....................
      END
C...............................................
C    STIFFNES MATRIX.
C...............................................
      SUBROUTINE  STIFFT (NE, NPE, GAUSS, WT, PD, ELSTIF,
     1                   PE, NOD, X, Y, IR,
     2                    IF, NELM, NNOD, NSIZ, NSTF)
      IMPLICIT REAL*8 (A-H,O-Z)
      DIMENSION NOD (NELM,9), X (NNOD), Y(NNOD),
     1          GAUSS (7,7), WT(7,7), ELSTIF(18,18),
     2          AKC (9,9), AKV (9,9), AKW (9,9), AKWD(9,9),
     3          DSIK(9), DSIE(9), SI(9), DSIKM(9), DSIEM(9), SIM(9),
     4          DSKK(9), DSEE(9), DSKE(9), XJ(9), YJ(9), ,
     6          AJI(2,2), AJ(2,2), DSIK1(9), DSIE1(9)
c    7           VPROP (15)
C..................................................................
      CALL ARR2ZF (AKC, 9)
      CALL ARR2ZF (AKV, 9)
```

```
      CALL ARR2ZF (AKW, 9)
      CALL ARR2ZF (AKWD, 9)
      CALL ARR2ZF (ELSTIF, NSTF)
C..........
      DO I=1,NPE
         XJ(I)=X(NOD(NE,I))
         YJ(I)=Y(NOD(NE,I))
      ENDDO

      DO 14 KI=1,IF
         AKESI=GAUSS(KI,IF)
         DO 14 KJ=1,IF
            ETA=GAUSS(KJ,IF)
            CALL SHAPEB  (AKESI, ETA, DSIK1, DSIE1, SI, NPE,
              1 DSIKM, DSIEM, SIM, ELLGTH, NE, NOD, X, Y, NELM, NNOD,
     2  DSIK, DSIE, PD, PE)

            CALL JACOB2 (AJ, AJI, DET, XJ, YJ, DSIK, DSIE,
     1                       NPE, NE)

          Da=PD
            COEF= DET*WT(KI,IF)*WT(KJ,IF)

              DO 16 M=1,NPE

               DSXM= DSIKM(M) * AJI(1,1) + DSIEM(M) * AJI(1,2)
               DSYM= DSIKM(M) * AJI(2,1) + DSIEM(M) * AJI(2,2)
               DO 16 N=1,NPE

      DSXN=  DSIK(N) * AJI(1,1) + DSIE(N) * AJI(1,2)
                  DSYN=  DSIK(N) * AJI(2,1) + DSIE(N) * AJI(2,2)

                  DSXN1= DSIK1(N) * AJI(1,1) + DSIE1(N) * AJI(1,2)
                  DSYN1= DSIK1(N) * AJI(2,1) + DSIE1(N) * AJI(2,2)

C     FOR CDR EQUATION POSITIVE REACTION TERM IS FOR EXPONENTIAL RGIME
AND
C     NEGATIVE IS FOR PRODUCTION.  INSTEAD OF Da WE USE PE*Da FOR CDR
EQUATION
       AKC(M,N)= AKC(M,N)+(DSXM*DSXN+DSYM*DSYN+ Da*SIM(M)*SI(N))*COEF
                AKV(M,N)= AKV(M,N)+PE*(SIM(M)*DSXN1+SIM(M)*DSYN1)*COEF

16 CONTINUE
14 CONTINUE

100 DO I=1,NPE

      DO J=1,NPE
```

For CDR equation the coefficient of the reaction term is $P_e^* D_a$. See Sec. 4.5.1.

```
        ELSTIF(I,J)= AKC(I,J)+AKV(I,J)
      ENDDO
    ENDDO
    RETURN
    END
C..............................................

      SUBROUTINE OUTPUT (NNM, TF, NSIZ)
      IMPLICIT REAL*8 (A-H,O-Z)
      DIMENSION TF  (NSIZ)
      WRITE (2, 5120)
            DO 24  I=1,NNM
        WRITE (2,5130) I, TF(I)
 24   CONTINUE

 5120 FORMAT (1X,//,' RESULT (NODE NO., TEMPERATURE)',/)
 5130 FORMAT (1X,I4,2X,7(D11.5,2X))
      RETURN
      END
C..............................................
C       A=0 SUBROUTINE  (2-D) (FLOAT).
C..............................................
      SUBROUTINE ARR2ZF (A, N)
      IMPLICIT REAL*8 (A-H,O-Z)
      DIMENSION A(N,N)
      DO I= 1,N
         DO J=1,N
            A(I,J)=0.0
         ENDDO
      ENDDO
      RETURN
      END
C..............................................
C     A=B SUBROUTINE  (FLOAT)                 .
C..............................................
      SUBROUTINE RSAVE (A, B, N)
      IMPLICIT REAL*8 (A-H,O-Z)
      DIMENSION A(N),B(N)
      DO I=1,N
         A(I)=B(I)
      ENDDO
      RETURN
      END
C..............................................
C     A=B SUBROUTINE  (INTEGER)               .
C..............................................
      SUBROUTINE RSAVI (IA, IB, N)
      DIMENSION IA(N),IB(N)
```

```
      DO I=1,N
         IA(I)=IB(I)
      ENDDO
      RETURN
      END
C..........................................
```

Calculation of the element length for bilinear elements

```
 SUBROUTINE ELMTLNTH (NE, NOD, X, Y, NELM, NNOD, ELLGTH,
      1          DSIX, DSIY)


   IMPLICIT REAL*8 (A-H,O-Z)
   DIMENSION  NOD (NELM,9), X (NNOD), Y(NNOD)
   DIMENSION  DSI(2), NP(2,4), NQ(4,2)
C............................................................
   NQ(1,2)=2
   NQ(1,1)=1
   NQ(2,2)=3
   NQ(2,1)=2
   NQ(3,2)=4
   NQ(3,1)=3
   NQ(4,2)=1
   NQ(4,1)=4
   DO I=1,4
      DO J=1,2
         NP (3-J,I) = NOD (NE, NQ (I,3-J))
      ENDDO
   ENDDO

         DSIX =0.0
         DSIY =0.0
         DSI(1)=0.5
         DSI(2)=0.5
         DO L=1,4
                DO I= 1,2
           DSIX = (X(NP(1,2))-X(NP(1,1))+X(NP(1,3))-X(NP(2,3)))/2
           DSIY = (Y(NP(2,2))-Y(NP(1,2))+Y(NP(1,4))-Y(NP(2,4)))/2
         ENDDO
         ENDDO

      ELLGTH = DSQRT(DSIX*DSIY)

  RETURN
  END
C............................................................
```

Calculation of shape funtions

```
 SUBROUTINE SHAPEB (AKESI, ETA, DSIK1, DSIE1, SI, NPE,
      1 DSIKM, DSIEM, SIM, ELLGTH, NE, NOD, X, Y, NELM, NNOD,
 2 DSIK, DSIE, PD,PE)
  IMPLICIT REAL*8 (A-H,O-Z)
  DIMENSION DSIK1(9),DSIE1(9),SI(9),DSIKM(9),DSIEM(9),SIM(9),X(NNOD)
  DIMENSION DSIK(9),DSIE(9)


  CALL ELMTLNTH (NE, NOD, X, Y, NELM, NNOD, ELLGTH,
      1         DSIX, DSIY)

  Da=PD

C***************************************************************************
**
```

Fifth-order RFB bubble function for DR equation

```
c BA=ELLGTH*(1+Da*(ELLGTH**2)*0.1667+(Da**2)*(ELLGTH**4)*0.0083)
c     BA=1/BA

c     A=-Da*(ELLGTH**3)*0.04167-(Da**2)*(ELLGTH**5)*0.002083
c     B=-Da*(ELLGTH**3)*0.02083-(Da**2)*(ELLGTH**5)*0.003125
c     C= (Da**2)*(ELLGTH**5)*0.0010417
c     D= (Da**2)*(ELLGTH**5)*0.0002604


c     SI(1)=0.25*(1-AKESI)*(1-ETA)+BA*(A*(1-AKESI**2)*(1-ETA**2)+
c  1  B*(1-AKESI**2)*(1-AKESI)*(1-ETA**2)*(1-ETA)+
c  2  C*((1-AKESI**2)**2)*((1-ETA**2)**2)+
c  3  D*(1-AKESI)*((1-AKESI**2)**2)*(1-ETA)*((1-ETA**2)**2))

c     SI(2)=0.25*(1+AKESI)*(1-ETA)+BA*(A*(1-AKESI**2)*(1-ETA**2)+
c  1  B*(1-AKESI**2)*(1+AKESI)*(1-ETA**2)*(1-ETA)+
c  2  C*((1-AKESI**2)**2)*((1-ETA**2)**2)+
c  3  D*(1+AKESI)*((1-AKESI**2)**2)*(1-ETA)*((1-ETA**2)**2))

c     SI(3)=0.25*(1+AKESI)*(1+ETA)+BA*(A*(1-AKESI**2)*(1-ETA**2)+
c  1  B*(1-AKESI**2)*(1+AKESI)*(1-ETA**2)*(1+ETA)+
c  2  C*((1-AKESI**2)**2)*((1-ETA**2)**2)+
c  3  D*(1+AKESI)*((1-AKESI**2)**2)*(1+ETA)*((1-ETA**2)**2))

c     SI(4)=0.25*(1-AKESI)*(1+ETA)+BA*(A*(1-AKESI**2)*(1-ETA**2)+
c  1  B*(1-AKESI**2)*(1-AKESI)*(1-ETA**2)*(1+ETA)+
c  2  C*((1-AKESI**2)**2)*((1-ETA**2)**2)+
c  3  D*(1-AKESI)*((1-AKESI**2)**2)*(1+ETA)*((1-ETA**2)**2))

c   DSIK(1)=-0.25*(1-ETA)
c   DSIK(2)= 0.25*(1-ETA)
```

```
c   DSIK(3)= 0.25*(1+ETA)
c   DSIK(4)=-0.25*(1+ETA)
C..............

c   DSIE(1)=-0.25*(1-AKESI)
c   DSIE(2)=-0.25*(1+AKESI)
c   DSIE(3)= 0.25*(1+AKESI)
c   DSIE(4)= 0.25*(1-AKESI)
c......................................................
```

Second-order STC bubble function for DR equation (see Eq. (3.102))

```
c          B=1/(1.6+16/(Da*(ELLGTH**2)))

c   DSIK(1)=-0.25*(1-ETA)
c   DSIK(2)= 0.25*(1-ETA)
c   DSIK(3)= 0.25*(1+ETA)
c   DSIK(4)=-0.25*(1+ETA)
C...............
c   DSIE(1)=-0.25*(1-AKESI)
c   DSIE(2)=-0.25*(1+AKESI)
c   DSIE(3)= 0.25*(1+AKESI)
c   DSIE(4)= 0.25*(1-AKESI)
C................
c   SI(1)=0.25*(1-AKESI)*(1-ETA)-B*(1-ETA**2)*(1-AKESI**2)
c   SI(2)=0.25*(1+AKESI)*(1-ETA)-B*(1-ETA**2)*(1-AKESI**2)
c   SI(3)=0.25*(1+AKESI)*(1+ETA)-B*(1-ETA**2)*(1-AKESI**2)
c   SI(4)=0.25*(1-AKESI)*(1+ETA)-B*(1-ETA**2)*(1-AKESI**2)

C***********************************************************************
**
```

Fourth-order RFB bubble function for CD equation

```
c  E=ELLGTH
c      PL=PE*E

c      B1=0.5*(PE**2)+(0.1667)*(PE**3)*E+(0.04167)*(PE**4)*(E**2)
c      B1=B1*(E**2)*0.25
c      B2=((0.1667)*(PE**3)+(0.04167)*(PE**4)*2*E)*(E*0.5)
c      B2=B2*(E**2)*0.25
c      B3=(0.04167)*(PE**4)*(E**2)*(0.25)
c      B3=B3*(E**2)*0.25
c      A=PL+(0.5)*(PL**2)+(0.1667)*(PL**3)+(0.04167)*(PL**4)
c      A=-1/A


c    DSIK1(1)=-0.25*(1-ETA)-A*(-2*B1*(AKESI**1)*(1-ETA**2)+
c 1    B2*(-(1-AKESI**2)-2*AKESI*(1-AKESI))*(1-ETA)*(1-ETA**2)-
c      2        4*B3*AKESI*(1-AKESI**2)*((1-ETA**2)**2))
```

```
c     DSIK1(2)= 0.25*(1-ETA)+A*(-2*B1*(AKESI**1)*(1-ETA**2)+
c       1       B2*((1-AKESI**2)-2*AKESI*(1+AKESI))*(1-ETA)*(1-ETA**2)-
c       2       4*B3*AKESI*(1-AKESI**2)*((1-ETA**2)**2))

c     DSIK1(3)= 0.25*(1+ETA)+A*(-2*B1*(AKESI**1)*(1-ETA**2)+
c       1       B2*((1-AKESI**2)-2*AKESI*(1+AKESI))*(1+ETA)*(1-ETA**2)-
c       2       4*B3*AKESI*(1-AKESI**2)*((1-ETA**2)**2))

c     DSIK1(4)=-0.25*(1+ETA)-A*(-2*B1*(AKESI**1)*(1-ETA**2)+
c       1       B2*(-(1-AKESI**2)-2*AKESI*(1-AKESI))*(1+ETA)*(1-ETA**2)-
c       2       4*B3*AKESI*(1-AKESI**2)*((1-ETA**2)**2))
c.............

c     DSIE1(1)=-0.25*(1-AKESI)-A*(-2*B1*(ETA**1)*(1-AKESI**2)+
c       1       B2*(-(1-ETA**2)-2*ETA*(1-ETA))*(1-AKESI)*(1-AKESI**2)-
c       2       4*B3*ETA*(1-ETA**2)*((1-AKESI**2)**2))

c     DSIE1(2)=-0.25*(1+AKESI)-A*(-2*B1*(ETA**1)*(1-AKESI**2)+
c       1       B2*(-(1-ETA**2)-2*ETA*(1-ETA))*(1+AKESI)*(1-AKESI**2)-
c       2       4*B3*ETA*(1-ETA**2)*((1-AKESI**2)**2))

c     DSIE1(3)= 0.25*(1+AKESI)+A*(-2*B1*(ETA**1)*(1-AKESI**2)+
c       1       B2*((1-ETA**2)-2*ETA*(1+ETA))*(1+AKESI)*(1-AKESI**2)-
c       2       4*B3*ETA*(1-ETA**2)*((1-AKESI**2)**2))

c     DSIE1(4)= 0.25*(1-AKESI)+A*(-2*B1*(ETA**1)*(1-AKESI**2)+
c       1       B2*((1-ETA**2)-2*ETA*(1+ETA))*(1-AKESI)*(1-AKESI**2)-
c       2       4*B3*ETA*(1-ETA**2)*((1-AKESI**2)**2))
c     DSIK(1)=-0.25*(1-ETA)
c     DSIK(2)= 0.25*(1-ETA)
c     DSIK(3)= 0.25*(1+ETA)
c     DSIK(4)=-0.25*(1+ETA)

C................
c     DSIE(1)=-0.25*(1-AKESI)
c     DSIE(2)=-0.25*(1+AKESI)
c     DSIE(3)= 0.25*(1+AKESI)
c     DSIE(4)= 0.25*(1-AKESI)

C-----------------------------------------------------------------------
```

nth-order STC bubble function for CD equation

```
c       N=10
c       M=N-1

c       PEL=PE*ELLGTH
c       B1=((2*N-1)*PEL)/(4*N*(N+1))
```

```
c     DSIK1(1)=-0.25*(1-ETA)-N*B1*(AKESI**M)*(1-ETA**N)
c     DSIK1(2)= 0.25*(1-ETA)+N*B1*(AKESI**M)*(1-ETA**N)
c     DSIK1(3)= 0.25*(1+ETA)+N*B1*(AKESI**M)*(1-ETA**N)
c     DSIK1(4)=-0.25*(1+ETA)-N*B1*(AKESI**M)*(1-ETA**N)
C.................
c     DSIE1(1)=-0.25*(1-AKESI)-N*B1*(ETA**M)*(1-AKESI**N)
c     DSIE1(2)=-0.25*(1+AKESI)-N*B1*(ETA**M)*(1-AKESI**N)
c     DSIE1(3)= 0.25*(1+AKESI)+N*B1*(ETA**M)*(1-AKESI**N)
c     DSIE1(4)= 0.25*(1-AKESI)+N*B1*(ETA**M)*(1-AKESI**N)

c     DSIK(1)=-0.25*(1-ETA)
c     DSIK(2)= 0.25*(1-ETA)
c     DSIK(3)= 0.25*(1+ETA)
c     DSIK(4)=-0.25*(1+ETA)
C..................
c     DSIE(1)=-0.25*(1-AKESI)
c     DSIE(2)=-0.25*(1+AKESI)
c     DSIE(3)= 0.25*(1+AKESI)
c     DSIE(4)= 0.25*(1-AKESI)

C----------------------------------------------------------------------------
```

Second-order STC bubble function for CDR equation

```
    E=ELLGTH
C--------------------- DISSIPATION --Da IS NEGATIVE------------

      ALF=0.5*((PE**2+4*PE*Da)**0.5)

C---------------------- PRODUCTION --Da IS POSITIVE------------

c     ALF=0.5*((ABS(PE**2+4*PE*DA))**0.5)

C----------------------------------------------------------------------------

 A1=0.5*PE+ALF
         A2=0.5*PE-ALF

      A=A1-A2
            B1=0.5*(A1**2-A2**2)

      ADI=(A+B1*E)*E
          B11=(B1*(E**2)*0.25)/(ADI)
          B12=-(B1*(E**2)*0.25)/(ADI)

     DSIK1(1)=-0.25*(1-ETA)-2*B11*AKESI*(1-ETA**2)
     DSIK1(2)= 0.25*(1-ETA)+2*B12*AKESI*(1-ETA**2)
     DSIK1(3)= 0.25*(1+ETA)+2*B12*AKESI*(1-ETA**2)
     DSIK1(4)=-0.25*(1+ETA)-2*B11*AKESI*(1-ETA**2)
```

```
c................................

      DSIE1(1)=-0.25*(1-AKESI)-2*B11*ETA*(1-AKESI**2)
      DSIE1(2)=-0.25*(1+AKESI)-2*B12*ETA*(1-AKESI**2)
      DSIE1(3)= 0.25*(1+AKESI)+2*B12*ETA*(1-AKESI**2)
      DSIE1(4)= 0.25*(1-AKESI)+2*B11*ETA*(1-AKESI**2)

      DSIK(1)=-0.25*(1-ETA)
      DSIK(2)= 0.25*(1-ETA)
      DSIK(3)= 0.25*(1+ETA)
      DSIK(4)=-0.25*(1+ETA)
c................................
      DSIE(1)=-0.25*(1-AKESI)
      DSIE(2)=-0.25*(1+AKESI)
      DSIE(3)= 0.25*(1+AKESI)
      DSIE(4)= 0.25*(1-AKESI)
c................................

      SI(1)=0.25*(1-AKESI)*(1-ETA)+B11*(1-ETA**2)*(1-AKESI**2)
      SI(2)=0.25*(1+AKESI)*(1-ETA)+B12*(1-ETA**2)*(1-AKESI**2)
      SI(3)=0.25*(1+AKESI)*(1+ETA)+B12*(1-ETA**2)*(1-AKESI**2)
      SI(4)=0.25*(1-AKESI)*(1+ETA)+B11*(1-ETA**2)*(1-AKESI**2)

c--------weight functions

    IF (NPE.EQ.4) THEN

      DSIKM(1)=-0.25*(1-ETA)
      DSIKM(2)= 0.25*(1-ETA)
      DSIKM(3)= 0.25*(1+ETA)
      DSIKM(4)=-0.25*(1+ETA)
C................................
      DSIEM(1)=-0.25*(1-AKESI)
      DSIEM(2)=-0.25*(1+AKESI)
      DSIEM(3)= 0.25*(1+AKESI)
      DSIEM(4)= 0.25*(1-AKESI)
C................................
      SIM(1)=0.25*(1-AKESI)*(1-ETA)
      SIM(2)=0.25*(1+AKESI)*(1-ETA)
      SIM(3)=0.25*(1+AKESI)*(1+ETA)
      SIM(4)=0.25*(1-AKESI)*(1+ETA)

    ELSEIF (NPE.EQ.8) THEN

      DSIK(1)=0.5*AKESI-0.5*AKESI*ETA-0.25*ETA**2+0.25*ETA
      DSIK(2)=0.5*AKESI-0.5*AKESI*ETA+0.25*ETA**2-0.25*ETA
      DSIK(3)=0.5*AKESI+0.5*AKESI*ETA+0.25*ETA**2+0.25*ETA
      DSIK(4)=0.5*AKESI+0.5*AKESI*ETA-0.25*ETA**2-0.25*ETA
      DSIK(5)=AKESI*(-1+ETA)
      DSIK(6)=0.5-0.5*ETA**2
```

```
      DSIK(7)=-AKESI*(1+ETA)
      DSIK(8)=-0.5+0.5*ETA**2
C................................
      DSIE(1)=0.5*ETA-0.25*AKESI**2-0.5*AKESI*ETA+0.25*AKESI
      DSIE(2)=0.5*ETA-0.25*AKESI**2+0.5*ETA*AKESI-0.25*AKESI
      DSIE(3)=0.5*ETA+0.25*AKESI**2+0.5*AKESI*ETA+0.25*AKESI
      DSIE(4)=0.5*ETA+0.25*AKESI**2-0.5*AKESI*ETA-0.25*AKESI
      DSIE(5)=-0.5+0.5*AKESI**2
      DSIE(6)=-(1+AKESI)*ETA
      DSIE(7)=0.5-0.5*AKESI**2
      DSIE(8)=(-1+AKESI)*ETA
C...............................
      SI(1)=0.25*(1-AKESI)*(1-ETA)*(-1-AKESI-ETA)
      SI(2)=0.25*(1+AKESI)*(1-ETA)*(-1+AKESI-ETA)
      SI(3)=0.25*(1+AKESI)*(1+ETA)*(-1+AKESI+ETA)
      SI(4)=0.25*(1-AKESI)*(1+ETA)*(-1-AKESI+ETA)
      SI(5)=0.5*(1-AKESI**2)*(1-ETA)
      SI(6)=0.5*(1+AKESI)*(1-ETA**2)
      SI(7)=0.5*(1-AKESI**2)*(1+ETA)
      SI(8)=0.5*(1-AKESI)*(1-ETA**2)
    ELSEIF (NPE.EQ.9) THEN
      DSIK(1)=0.25*(2*AKESI-1)*(ETA**2-ETA)
      DSIK(2)=0.25*(2*AKESI+1)*(ETA**2-ETA)
      DSIK(3)=0.25*(2*AKESI+1)*(ETA**2+ETA)
      DSIK(4)=0.25*(2*AKESI-1)*(ETA**2+ETA)
      DSIK(5)=-AKESI*(ETA**2-ETA)
      DSIK(6)=0.5*(2*AKESI+1)*(1-ETA**2)
      DSIK(7)=-AKESI*(ETA**2+ETA)
      DSIK(8)=0.5*(2*AKESI-1)*(1-ETA**2)
      DSIK(9)=-2*AKESI*(1-ETA**2)
C................................
      DSIE(1)=0.25*(AKESI**2-AKESI)*(2*ETA-1)
      DSIE(2)=0.25*(AKESI**2+AKESI)*(2*ETA-1)
      DSIE(3)=0.25*(AKESI**2+AKESI)*(2*ETA+1)
      DSIE(4)=0.25*(AKESI**2-AKESI)*(2*ETA+1)
      DSIE(5)=0.5*(1-AKESI**2)*(2*ETA-1)
      DSIE(6)=-ETA*(AKESI**2+AKESI)
      DSIE(7)=0.5*(1-AKESI**2)*(2*ETA+1)
      DSIE(8)=-ETA*(AKESI**2-AKESI)
      DSIE(9)=-2*ETA*(1-AKESI**2)
C.................................
      SI(1)=0.25*(AKESI**2-AKESI)*(ETA**2-ETA)
      SI(2)=0.25*(AKESI**2+AKESI)*(ETA**2-ETA)
      SI(3)=0.25*(AKESI**2+AKESI)*(ETA**2+ETA)
      SI(4)=0.25*(AKESI**2-AKESI)*(ETA**2+ETA)
      SI(5)=0.5*(1-AKESI**2)*(ETA**2-ETA)
      SI(6)=0.5*(AKESI**2+AKESI)*(1-ETA**2)
      SI(7)=0.5*(1-AKESI**2)*(ETA**2+ETA)
      SI(8)=0.5*(AKESI**2-AKESI)*(1-ETA**2)
      SI(9)=(1-AKESI**2)*(1-ETA**2)
```

```
   ENDIF
   RETURN
   END
C..................................................
C  CALCULATION OF JACOBIAN
C..................................................
      SUBROUTINE JACOB2 (AJ, AJI, DET, X, Y,
     1                   DSIK, DSIE, N, NE)
     IMPLICIT REAL*8 (A-H,O-Z)
     DIMENSION AJ (2,2), AJI (2,2),
     1   X(N), Y(N),
     2   DSIK(N), DSIE(N)

     DO I=1,2
      DO J=1,2
       AJ  (I,J)  = 0.0
       AJI (I,J)  = 0.0
      ENDDO
     ENDDO
     DO I=1,N
      AJ(1,1) = AJ(1,1) + X(I) * DSIK(I)
      AJ(1,2) = AJ(1,2) + Y(I) * DSIK(I)
      AJ(2,1) = AJ(2,1) + X(I) * DSIE(I)
      AJ(2,2) = AJ(2,2) + Y(I) * DSIE(I)
     ENDDO
     DET= AJ(1,1)*AJ(2,2)-AJ(1,2)*AJ(2,1)

     IF (DET.LE.0.0) THEN
       WRITE (2,110) NE,DET
 110    FORMAT  (1X,'ERROR : ZERRO OR NEGATIVE JACOBIAN=',I6,G20.5)
       STOP
     ENDIF

     AJI(1,1) =  AJ(2,2) / DET
     AJI(1,2) = -AJ(1,2) / DET
     AJI(2,1) = -AJ(2,1) / DET
     AJI(2,2) =  AJ(1,1) / DET

     RETURN
     END
C..............................................

     SUBROUTINE ARRZRI (IA , N)
     DIMENSION IA(N)
     DO I= 1,N
      IA(I)=0
     ENDDO
     RETURN
     END
C..................................................
```

```
C    CALCULATION OF ERROR NORM                                  .
C.............................................................
     SUBROUTINE VNORM (VN , NEQ , G1 , G2 , NSIZ)
     IMPLICIT REAL*8 (A-H,O-Z)
     DIMENSION G2(NSIZ),G1(NSIZ)
     A=0.0
     B=0.0
     DO I=1,NEQ
      A=A+(G2(I)-G1(I))**2
      B=B+G2(I)**2
     ENDDO
     IF (A.LT.1.00D-10.AND.B.LT.1.00D-10) THEN
       VN =0.0
     ELSE
       VN=DSQRT(A)/DSQRT(B)
     ENDIF
     RETURN
     END

C................................................
C    RENEWAL SUBROUTINE  WITH OVER-RELAXATION   .
C................................................
     SUBROUTINE RENEWAL (A , B , N , OMEGA)
     IMPLICIT REAL*8 (A-H,O-Z)
     DIMENSION A(N),B(N)
     DO I=1,N
      A(I)=A(I)+(B(I)-A(I))*OMEGA
     ENDDO
     RETURN
     END
C............................................................
C     FRONTAL SOLVER                                         .
C............................................................
     SUBROUTINE FRONT
    1    (AA, RR, IEL, NOP, MAXEL, MAXST, LDEST, NK, MAXFR, EQ,
    2     LHED, LPIV, JMOD, QQ, PVKOL, DIS, R1, NCOD, BC, NOPP,
    3     MDF, NDN, MAXDF, NEL, MAXTE, NTOV, LCOL, NELL, NPE)
C
C    FRONTAL ELIMINATION ROUTINE USING DIAGONAL PIVOTING
C
     IMPLICIT DOUBLE PRECISION(A-H,O-Z)
     DIMENSION AA   (MAXST,MAXST), RR   (MAXST)
     DIMENSION NOP  (MAXEL, 9)
     DIMENSION LDEST(MAXST), NK   (MAXST)
     DIMENSION EQ   (MAXFR, MAXFR), LHED (MAXFR)
     DIMENSION LPIV (MAXFR)
     DIMENSION JMOD (MAXFR), QQ   (MAXFR), PVKOL(MAXFR)
     DIMENSION DIS  (MAXTE), R1   (MAXDF), NCOD (MAXDF)
     DIMENSION BC   (MAXDF), NOPP (MAXDF), MDF  (MAXDF)
     DIMENSION NDN  (MAXDF)
```

```
C
      NLP=6
      ND1=14
C
C     ........
C
      NMAX=MAXFR
      NCRIT=50
      NLARG=MAXFR-10
      IF(IEL.EQ.1) NELL = 0
C   *************************
      IF(IEL.GT.1) GO TO 18
      LCOL = 0
      DO 16 I = 1,NMAX
      DO 16 J = 1,NMAX
      EQ(J,I) = 0.
   16 CONTINUE
   18 NELL = NELL+1
      N = NELL
      JDN = NDN(NELL)
      KC = 0
      DO 22 J = 1,NPE
      NN = NOP(N,J)
      M = IABS(NN)
      K = NOPP(M)
      IDF = MDF(M)
C *********************
      NR = (M - 1) * IDF
C *********************
C     R1(M) = RR(J)+R1(M)
C *********************
     DO 22 L = 1,IDF
C *********************
     NR=NR+1
     NL=(J-1)*IDF+L
     R1(NR)=R1(NR)+RR (NL)
C**********************
     KC = KC+1
     II = K+L-1
     IF(NN.LT.0)II = -II
     NK(KC) = II
    22 CONTINUE
C
C     SET UP HEADING VECTORS
C
    DO 36 LK = 1,KC
    NODE = NK(LK)
    IF(LCOL.EQ.0)GO TO 28
    DO 24 L = 1,LCOL
    LL = L
```

```
      IF(IABS(NODE).EQ.IABS(LHED(L)))GO TO 32
   24 CONTINUE
   28 LCOL = LCOL+1
      LDEST(LK) = LCOL
      LHED(LCOL) = NODE
      GO TO 36
   32 LDEST(LK) = LL
      LHED(LL) = NODE
   36 CONTINUE
      IF(LCOL.LE.NMAX)GO TO 54
      NERROR = 2
      WRITE(NLP,417)NERROR
      STOP
   54 CONTINUE
      DO 56 L = 1,KC
      LL = LDEST(L)
      DO 56 K = 1,KC
      KK = LDEST(K)
      EQ(KK,LL) = EQ(KK,LL)+AA(K,L)
   56 CONTINUE
      IF(LCOL.LT.NCRIT.AND.NELL.LT.NEL) RETURN
C
C   FIND OUT WHICH MATRIX ELEMENTS ARE FULLY ASSEMBLED
C
   60 LC = 0
      IR = 0
      DO 64 L = 1,LCOL
      KT = LHED(L)
      IF(KT.GE.0)GO TO 64
      LC = LC+1
      LPIV(LC) = L
      KRO = IABS(KT)
      IF(NCOD(KRO).NE.1)GO TO 64
      IR = IR+1
      JMOD(IR) = L
      NCOD(KRO) = 2
      R1(KRO) = BC(KRO)
   64 CONTINUE
C
C   MODIFY EQUATIONS WITH APPLIED BOUNDARY CONDITIONS
C
      IF(IR.EQ.0)GO TO 71
      DO 70 IRR = 1,IR
      K = JMOD(IRR)
      KH = IABS(LHED(K))
      DO 69 L = 1,LCOL
      EQ(K,L) = 0.
      LH = IABS(LHED(L))
      IF(LH.EQ.KH)EQ(K,L) = 1.
   69 CONTINUE
```

```
   70 CONTINUE
   71 CONTINUE
      IF(LC.GT.0)GO TO 72
      NCRIT = NCRIT+10
C      WRITE(NLP,484)NCRIT
      IF(NCRIT.LE.NLARG) RETURN
      NERROR = 3
      WRITE(NLP,418)NERROR
      STOP
   72 CONTINUE
C
C     SEARCH FOR ABSOLUTE PIVOT
C
      PIVOT = 0.
      DO 76 L = 1,LC
      LPIVC = LPIV(L)
      KPIVR = LPIVC
      PIVA = EQ(KPIVR,LPIVC)
      IF(ABS(PIVA).LT.ABS(PIVOT))GO TO 74
      PIVOT = PIVA
      LPIVCO = LPIVC
      KPIVRO = KPIVR
   74 CONTINUE
   76 CONTINUE
      IF(PIVOT.EQ.0.0)  RETURN
C
C     NORMALISE PIVOTAL ROW
C
      LCO = IABS(LHED(LPIVCO))
      KRO = LCO
C     IF(NIT.EQ.0.OR.NPRA.EQ.0)GO TO 78
C     WRITE(NLP,452)KRO,LCO,PIVOT
C  78 CONTINUE
      IF(ABS(PIVOT).LT.0.1D-08) WRITE(NLP,476)
      DO 80 L = 1,LCOL
      QQ(L) = EQ(KPIVRO,L)/PIVOT
   80 CONTINUE
      RHS = R1(KRO)/PIVOT
      R1(KRO) = RHS
      PVKOL(KPIVRO) = PIVOT
C
C     ELIMINATE THEN DELETE PIVOTAL ROW AND COLUMN
C
      IF(KPIVRO.EQ.1)GO TO 104
      KPIVR = KPIVRO-1
      DO 100 K = 1,KPIVR
      KRW = IABS(LHED(K))
      FAC = EQ(K,LPIVCO)
      PVKOL(K) = FAC
      IF(LPIVCO.EQ.1.OR.FAC.EQ.0.)GO TO 88
```

```
      LPIVC = LPIVCO-1
      DO 84 L = 1,LPIVC
      EQ(K,L) = EQ(K,L)-FAC*QQ(L)
   84 CONTINUE
   88 IF(LPIVCO.EQ.LCOL)GO TO 96
       LPIVC = LPIVCO+1
       DO 92 L = LPIVC,LCOL
       EQ(K,L-1) = EQ(K,L)-FAC*QQ(L)
   92 CONTINUE
   96 R1(KRW) = R1(KRW)-FAC*RHS
  100 CONTINUE
  104 IF(KPIVRO.EQ.LCOL)GO TO 128
      KPIVR = KPIVRO+1
      DO 124 K = KPIVR,LCOL
      KRW = IABS(LHED(K))
      FAC = EQ(K,LPIVCO)
      PVKOL(K) = FAC
      IF(LPIVCO.EQ.1)GO TO 112
      LPIVC = LPIVCO-1
      DO 108 L = 1,LPIVC
      EQ(K-1,L) = EQ(K,L)-FAC*QQ(L)
  108 CONTINUE
  112 IF(LPIVCO.EQ.LCOL)GO TO 120
      LPIVC = LPIVCO+1
      DO 116 L = LPIVC,LCOL
      EQ(K-1,L-1) = EQ(K,L)-FAC*QQ(L)
  116 CONTINUE
  120 R1(KRW) = R1(KRW)-FAC*RHS
  124 CONTINUE
  128 CONTINUE
C
C   WRITE PIVOTAL EQUATION ON DISC
C
      WRITE(ND1) KRO,LCOL,LPIVCO,(LHED(L),QQ(L),L = 1,LCOL)
      DO 130 L = 1,LCOL
      EQ(L,LCOL) = 0.
      EQ(LCOL,L) = 0.
  130 CONTINUE
C
C   REARRANGE HEADING VECTORS
C
      LCOL = LCOL-1
      IF(LPIVCO.EQ.LCOL+1)GO TO 136
      DO 132 L = LPIVCO,LCOL
      LHED(L) = LHED(L+1)
  132 CONTINUE
  136 CONTINUE
C
C   DETERMINE WHETHER TO ASSEMBLE,ELIMINATE,OR BACKSUBSTITUTE
C
```

```
      IF(LCOL.GT.NCRIT)GO TO 60
      IF(NELL.LT.NEL) RETURN
      IF(LCOL.GT.1)GO TO 60
      LCO = IABS(LHED(1))
      KPIVRO = 1
      PIVOT = EQ(1,1)
      KRO = LCO
      LPIVCO = 1
      QQ(1) = 1.
C     IF(NIT.EQ.0.OR.NPRA.EQ.0)GO TO 148
C     WRITE(NLP,452)LCO,KRO,PIVOT
      IF(ABS(PIVOT).LT.1D-08)GO TO 152
C 148 CONTINUE
      R1(KRO) = R1(KRO)/PIVOT
      WRITE(ND1) KRO,LCOL,LPIVCO,LHED(1),QQ(1)
C
C *** START BACK-SUBSTITUTION
C
      CALL BACSUB
     1    (NTOV ,NCOD ,BC   ,R1   ,DIS  ,MAXFR,QQ   ,LHED ,ND1 )

C *** MAIN EXIT WITH SOLUTION ******************************************
C
  152 CONTINUE
  417 FORMAT(/'  NERROR=',I5//
     1 '  THE DIFFERENCE NMAX-NCRIT IS NOT SUFFICIENTLY LARGE'
     1/'  TO PERMIT THE ASSEMBLY OF THE NEXT ELEMENT---'
     1/'  EITHER INCREASE NMAX OR LOWER NCRIT'
     1/)
  418 FORMAT(/'  NERROR=',I5//
     1 '  THERE ARE NO MORE ROWS FULLY SUMMED,THIS MAY BE DUE TO---'
     1/'  (1)INCORRECT CODING OF NOP OR NK ARRAYS'
     1/'  (2)INCORRECT VALUE OF NCRIT. INCREASE NCRIT TO PERMIT'
     1/'      WHOLE FRONT TO BE ASSEMBLED'
     1/)
C 452 FORMAT(13H PIVOTAL ROW=,I4,16H PIVOTAL COLUMN=,I4,7H PIVOT=,E20.10
C   1)
  476 FORMAT('  WARNING-MATRIX SINGULAR OR ILL CONDITIONED')
  484 FORMAT('  FRONTWIDTH VALUE=',I4)
      RETURN
      END
C
C **********************************************************************
C
      SUBROUTINE BACSUB
     1 (NTOTL,IFIX ,VFIX ,RHS  ,SOLN ,MFRNT,RWORK,IWORK,IDV2)
C
C **********************************************************************
```

```
C
   IMPLICIT DOUBLE PRECISION(A-H,O-Z)
   DIMENSION IFIX (NTOTL) ,VFIX (NTOTL) ,RHS  (NTOTL) ,SOLN (NTOTL)
   DIMENSION RWORK(MFRNT) ,IWORK(MFRNT)
C
C ***
C
C ***
          DO 4990 IPOS=1,NTOTL
          SOLN(IPOS)=0.0
          IF(IFIX(IPOS).NE.0) SOLN(IPOS)=VFIX(IPOS)
 4990       CONTINUE
C
   DO 5000 KPOS=1,NTOTL
C
   BACKSPACE IDV2
   READ(IDV2)     IPOS,IFRNT,JFRNT,(IWORK(K),RWORK(K),K=1,IFRNT)
   BACKSPACE IDV2
C
   IF(IFIX(IPOS).NE.0)  GO TO 5000
C
   WW            = 0.0
   RWORK(JFRNT) = 0.0
C
          DO 5010 K=1,IFRNT
          JPOS=IABS(IWORK(K))
          WW  =WW - RWORK(K)*SOLN(JPOS)
5010      CONTINUE
C
   SOLN(IPOS)=RHS(IPOS)+WW
5000 CONTINUE
C
C ***
   RETURN
   END
C..........................................
C     PRE-FRONT ROUTINE                     .
C..........................................
   SUBROUTINE PREFNT  (NNM    , NEL    , NOP  , NPE  , MAXEL)
   DIMENSION NOP  (MAXEL , 9)
   NLAST = 0
   DO 12 I = 1,NNM
   DO 8  N = 1,NEL
C   JDN = NDN(N)
   DO 4 L =  1,NPE
   IF(NOP(N,L).NE.I)GO TO 4
   NLAST1 = N
   NLAST = N
```

```
   L1 = L
 4 CONTINUE
 8 CONTINUE
   IF(NLAST.EQ.0) GO TO 12
   NOP(NLAST,L1) = -NOP(NLAST,L1)
   NLAST = 0
12 CONTINUE
   RETURN
   END
```

6.3 Input File Structure

The sample input file for mesh scheme BLNR1 and boundary conditions BC1 and BC2. BLNR1 BC2

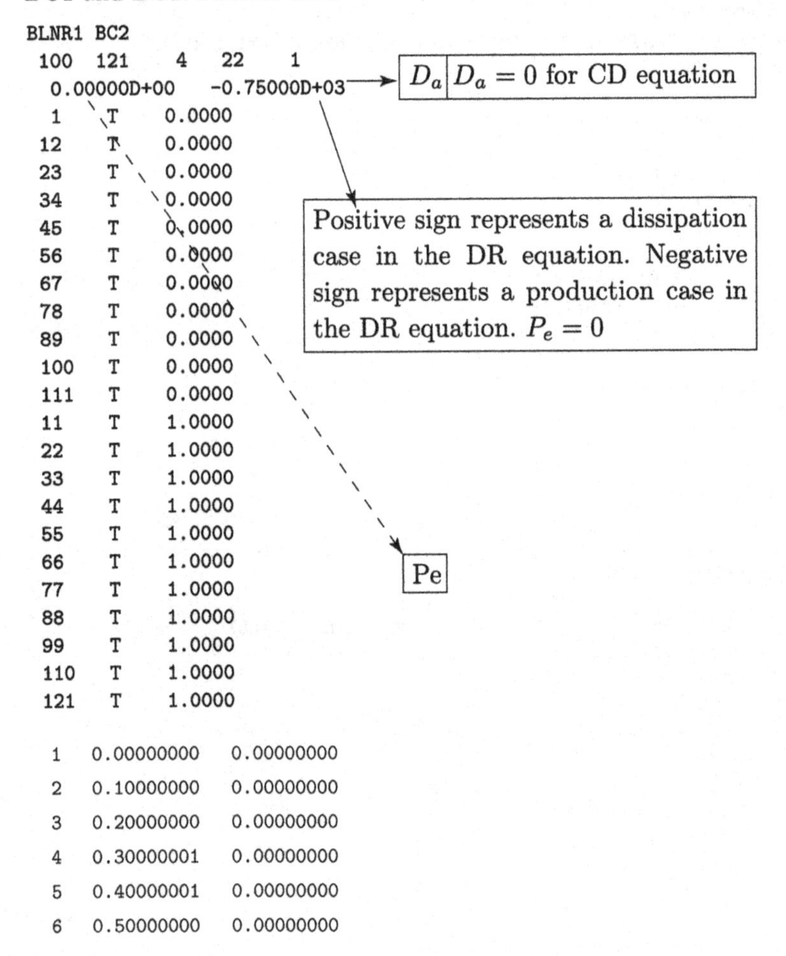

```
BLNR1 BC2
 100  121    4   22    1
  0.00000D+00    -0.75000D+03
 1    T     0.0000
 12   T     0.0000
 23   T     0.0000
 34   T     0.0000
 45   T     0.0000
 56   T     0.0000
 67   T     0.0000
 78   T     0.0000
 89   T     0.0000
 100  T     0.0000
 111  T     0.0000
 11   T     1.0000
 22   T     1.0000
 33   T     1.0000
 44   T     1.0000
 55   T     1.0000
 66   T     1.0000
 77   T     1.0000
 88   T     1.0000
 99   T     1.0000
 110  T     1.0000
 121  T     1.0000

 1   0.00000000   0.00000000
 2   0.10000000   0.00000000
 3   0.20000000   0.00000000
 4   0.30000001   0.00000000
 5   0.40000001   0.00000000
 6   0.50000000   0.00000000
```

```
 7   0.60000002   0.00000000
 8   0.70000005   0.00000000
 9   0.80000007   0.00000000
10   0.90000010   0.00000000
11   1.00000012   0.00000000
12   0.00000000   0.10000000
13   0.10000000   0.10000000
14   0.20000000   0.10000000
15   0.30000001   0.10000000
16   0.40000001   0.10000000
17   0.50000000   0.10000000
18   0.60000002   0.10000000
19   0.70000005   0.10000000
20   0.80000007   0.10000000
21   0.90000010   0.10000000
22   1.00000012   0.10000000
23   0.00000000   0.20000000
24   0.10000000   0.20000000
25   0.20000000   0.20000000
26   0.30000001   0.20000000
27   0.40000001   0.20000000
28   0.50000000   0.20000000
29   0.60000002   0.20000000
30   0.70000005   0.20000000
31   0.80000007   0.20000000
32   0.90000010   0.20000000
33   1.00000012   0.20000000
34   0.00000000   0.30000001
35   0.10000000   0.30000001
36   0.20000000   0.30000001
37   0.30000001   0.30000001
38   0.40000001   0.30000001
39   0.50000000   0.30000001
40   0.60000002   0.30000001
41   0.70000005   0.30000001
42   0.80000007   0.30000001
43   0.90000010   0.30000001
44   1.00000012   0.30000001
45   0.00000000   0.40000001
46   0.10000000   0.40000001
47   0.20000000   0.40000001
48   0.30000001   0.40000001
49   0.40000001   0.40000001
50   0.50000000   0.40000001
51   0.60000002   0.40000001
52   0.70000005   0.40000001
53   0.80000007   0.40000001
54   0.40000001   0.40000001
55   1.00000012   0.40000001
```

Nodal coordinates

```
56   0.00000000   0.50000000
57   0.10000000   0.50000000
58   0.20000000   0.50000000
59   0.30000001   0.50000000
60   0.40000001   0.50000000
61   0.50000000   0.50000000
62   0.60000002   0.50000000
63   0.70000005   0.50000000
64   0.80000007   0.50000000
65   0.90000010   0.50000000
66   1.00000012   0.50000000
67   0.00000000   0.60000002
68   0.10000000   0.60000002
69   0.20000000   0.60000002
70   0.30000001   0.60000002
71   0.40000001   0.60000002
72   0.50000000   0.60000002
73   0.60000002   0.60000002
74   0.70000005   0.60000002
75   0.80000007   0.60000002
76   0.90000010   0.60000002
77   1.00000012   0.60000002
78   0.00000000   0.70000005
79   0.10000000   0.70000005
80   0.20000000   0.70000005
81   0.30000001   0.70000005
82   0.40000001   0.70000005
83   0.50000000   0.70000005
84   0.60000002   0.70000005
85   0.70000005   0.70000005
86   0.80000007   0.70000005
87   0.90000010   0.70000005
88   1.00000012   0.70000005
89   0.00000000   0.80000007
90   0.10000000   0.80000007
91   0.20000000   0.80000007
92   0.30000001   0.80000007
93   0.40000001   0.80000007
94   0.50000000   0.80000007
95   0.60000002   0.80000007
96   0.70000005   0.80000007
97   0.80000007   0.80000007
98   0.90000010   0.80000007
99   1.00000012   0.80000007
100  0.00000000   0.90000010
101  0.10000000   0.90000010
```

For mesh scheme BLNR1 and boundary conditions BC1 boundary nodes are:

```
100 121 4 40    1
  0.10000D+02  0.10000D+02
2      T      0.0000
3      T      0.0000
4      T      0.0000
5      T      0.0000
6      T      0.0000
7      T      0.0000
8      T      0.0000
9      T      0.0000
10     T      0.0000
1     T      0.0000
12     T      0.0000
23     T      0.0000
34     T      0.0000
45     T      0.0000
56     T      0.0000
67     T      0.0000
78     T      0.0000
89     T      0.0000
100    T      0.0000
111    T      0.0000
11     T      1.0000
22     T      1.0000
33     T      1.0000
44     T      1.0000
55     T      1.0000
66     T      1.0000
77     T      1.0000
88     T      1.0000
99     T      1.0000
110    T      1.0000
121    T      1.0000
112    T      1.0000
113    T      1.0000
114    T      1.0000
115    T      1.0000
116    T      1.0000
117    T      1.0000
118    T      1.0000
119    T      1.0000
120    T      1.0000
```

```
102          0.20000000          0.90000010
103          0.30000001          0.90000010
104          0.40000001          0.90000010
105          0.50000000          0.90000010
106          0.60000002          0.90000010
107          0.70000005          0.90000010
108          0.80000007          0.90000010
109          0.90000010          0.90000010
110          1.00000012          0.90000010
111          0.00000000          1.00000012
112          0.10000000          1.00000012
113          0.20000000          1.00000012
114          0.30000001          1.00000012
115          0.40000001          1.00000012
116          0.50000000          1.00000012
117          0.60000002          1.00000012
118          0.70000005          1.00000012
119          0.80000007          1.00000012
120          0.90000010          1.00000012
121          1.00000012          1.00000012
  1    1    2   13   12
  2    2    3   14   13
  3    3    4   15   14
  4    4    5   16   15
  5    5    6   17   16
  6    6    7   18   17
  7    7    8   19   18
  8    8    9   20   19
  9    9   10   21   20
 10   10   11   22   21
 11   12   13   24   23
 12   13   14   25   24
 13   14   15   26   25
 14   15   16   27   26
 15   16   17   28   27
 16   17   18   29   28
 17   18   19   30   29
 18   19   20   31   30
 19   20   21   32   31
 20   21   22   33   32
 21   23   24   35   34
 22   24   25   36   35
 23   25   26   37   36
 24   26   27   38   37
 25   27   28   39   38
 26   28   29   40   39
 27   29   30   41   40
 28   30   31   42   41
```

```
29 31 32 43 42
30 32 33 44 43
31 34 35 46 45
32 35 36 47 46
33 36 37 48 47
34 37 38 49 48
35 38 39 50 49
36 39 40 51 50
37 40 41 52 51
38 41 42 53 52
39 42 43 54 53
40 43 44 55 54
41 45 46 57 56
42 46 47 58 57
43 47 48 59 58
44 48 49 60 59
45 49 50 61 60
46 50 51 62 61
47 51 52 63 62
48 52 53 64 63
49 53 54 65 64
50 54 55 66 65
51 56 57 68 67
52 57 58 69 68
53 58 59 70 69
54 59 60 71 70
55 60 61 72 71
56 61 62 73 72
57 62 63 74 73
58 63 64 75 74
59 64 65 76 75
60 65 66 77 76
61 67 68 79 78
62 68 69 80 79
63 69 70 81 80
64 70 71 82 81
65 71 72 83 82
66 72 73 84 83
67 73 74 85 84
68 74 75 86 85
69 75 76 87 86
70 76 77 88 87
71 78 79 90 89
72 79 80 91 90
73 80 81 92 91
74 81 82 93 92
75 82 83 94 93
76 83 84 95 94
77 84 85 96 95
78 85 86 97 96
```

Element connectivity data

```
 79   86   87   98   97
 80   87   88   99   98
 81   89   90  101  100
 82   90   91  102  101
 83   91   92  103  102
 84   92   93  104  103
 85   93   94  105  104
 86   94   95  106  105
 87   95   96  107  106
 88   96   97  108  107
 89   97   98  109  108
 90   98   99  110  109
 91  100  101  112  111
 92  101  102  113  112
 93  102  103  114  113
 94  103  104  115  114
 95  104  105  116  115
 96  105  106  117  116
 97  106  107  118  117
 98  107  108  119  118
 99  108  109  120  119
100  109  110  121  120
```

6.4 Output Files

In this section a number of sample results (program outputs) for the DR, CD, and CDR equations are presented.

6.4.1 *DR equation—dissipation case*

Results corresponding to the fifth-order RFB bubble function-enriched elements and $D_a = 750$. Mesh scheme BLNR1 and boundary conditions BC1.

```
TITLE : FIELD VARIABLE
RESULT (NODE NO., FIELD VARIABLE)
     1 0.00000D+00
     2 0.00000D+00
     3 0.00000D+00
     4 0.00000D+00
     5 0.00000D+00
     6 0.00000D+00
     7 0.00000D+00
     8 0.00000D+00
     9 0.00000D+00
    10 0.00000D+00
    11 0.10000D+01
```

Results for the DR equation, dissipation case, $D_a = 750$, fifth-order RFB bubble function (see Fig. 4.3 and Sec. 3.3.1).

```
12  0.00000D+00
13  0.10073D-08
14  0.50231D-08
15  0.65378D-07
16  0.40206D-06
17  0.61789D-05
18  0.43321D-04
19  0.66872D-03
20  0.50890D-02
21  0.10974D+00
22  0.10000D+01
23  0.00000D+00
24  0.72873D-08
25  0.18940D-07
26  0.92558D-07
27  0.11428D-05
28  0.80250D-05
29  0.10967D-03
30  0.86407D-03
31  0.11973D-01
32  0.10840D+00
33  0.10000D+01
34  0.00000D+00
35  0.36465D-07
36  0.12463D-06
37  0.25281D-06
38  0.12873D-05
39  0.14051D-04
40  0.11332D-03
41  0.13029D-02
42  0.11789D-01
43  0.10885D+00
44  0.10000D+01
45  0.00000D+00
46  0.58811D-06
47  0.72859D-06
48  0.15456D-05
49  0.28542D-05
50  0.15120D-04
51  0.14286D-03
52  0.12850D-02
53  0.11849D-01
54  0.10883D+00
55  0.10000D+01
56  0.00000D+00
57  0.37283D-05
58  0.10325D-04
59  0.10512D-04
60  0.16300D-04
61  0.28707D-04
```

```
 62  0.15421D-03
 63  0.13003D-02
 64  0.11855D-01
 65  0.10884D+00
 66  0.10000D+01
 67  0.00000D+00
 68  0.56296D-04
 69  0.74254D-04
 70  0.12737D-03
 71  0.12591D-03
 72  0.15539D-03
 73  0.26592D-03
 74  0.14109D-02
 75  0.11923D-01
 76  0.10889D+00
 77  0.10000D+01
 78  0.00000D+00
 79  0.40193D-03
 80  0.10012D-02
 81  0.10449D-02
 82  0.12991D-02
 83  0.12967D-02
 84  0.14111D-02
 85  0.23475D-02
 86  0.12790D-01
 87  0.10925D+00
 88  0.10000D+01
 89  0.00000D+00
 90  0.61005D-02
 91  0.79897D-02
 92  0.11926D-01
 93  0.11814D-01
 94  0.11857D-01
 95  0.11922D-01
 96  0.12790D-01
 97  0.19962D-01
 98  0.11450D+00
 99  0.10000D+01
100  0.00000D+00
101  0.47138D-01
102  0.10941D+00
103  0.10859D+00
104  0.10885D+00
105  0.10884D+00
106  0.10889D+00
107  0.10925D+00
108  0.11450D+00
109  0.15688D+00
110  0.10000D+01
111  0.00000D+00
```

```
112  0.10000D+01
113  0.10000D+01
114  0.10000D+01
115  0.10000D+01
116  0.10000D+01
117  0.10000D+01
118  0.10000D+01
119  0.10000D+01
120  0.10000D+01
121  0.10000D+01
```

6.4.2 *DR equation—production case*

Results corresponding to the second-order STC bubble function and $D_a = -750$. Mesh scheme BLNR1 and boundary conditions BC2.

```
TITLE : FIELD VARIABLE

RESULT (NODE NO., FIELD VARIABLE)
 1 0.00000D+00
 2 0.46304D+00
 3 -.84438D+00
 4 0.10767D+01
 5 -.11190D+01
 6 0.96387D+00
 7 -.63862D+00
 8 0.20067D+00
 9 0.27269D+00
10 -.69792D+00
11 0.10000D+01
12 0.00000D+00
13 0.46305D+00
14 -.84438D+00
15 0.10767D+01
16 -.11190D+01
17 0.96386D+00
18 -.63861D+00
19 0.20067D+00
20 0.27269D+00
21 -.69793D+00
22 0.10000D+01
23 0.00000D+00
24 0.46305D+00
25 -.84438D+00
26 0.10767D+01
27 -.11190D+01
28 0.96386D+00
```

Results for the DR equation, production case, $D_a = -750$, using second-order STC bubble function.

```
29  -.63861D+00
30  0.20067D+00
31  0.27269D+00
32  -.69793D+00
33  0.10000D+01
34  0.00000D+00
35  0.46305D+00
36  -.84438D+00
37  0.10767D+01
38  -.11190D+01
39  0.96387D+00
40  -.63862D+00
41  0.20067D+00
42  0.27269D+00
43  -.69793D+00
44  0.10000D+01
45  0.00000D+00
46  0.46304D+00
47  -.84438D+00
48  0.10767D+01
49  -.11190D+01
50  0.96386D+00
51  -.63862D+00
52  0.20067D+00
53  0.27269D+00
54  -.69792D+00
55  0.10000D+01
56  0.00000D+00
57  0.46305D+00
58  -.84438D+00
59  0.10767D+01
60  -.11190D+01
61  0.96387D+00
62  -.63862D+00
63  0.20067D+00
64  0.27269D+00
65  -.69793D+00
66  0.10000D+01
67  0.00000D+00
68  0.46305D+00
69  -.84438D+00
70  0.10767D+01
71  -.11190D+01
72  0.96387D+00
73  -.63861D+00
74  0.20067D+00
75  0.27269D+00
76  -.69793D+00
77  0.10000D+01
78  0.00000D+00
```

```
 79  0.46305D+00
 80  -.84438D+00
 81  0.10767D+01
 82  -.11190D+01
 83  0.96387D+00
 84  -.63862D+00
 85  0.20067D+00
 86  0.27269D+00
 87  -.69793D+00
 88  0.10000D+01
 89  0.00000D+00
 90  0.46304D+00
 91  -.84438D+00
 92  0.10767D+01
 93  -.11190D+01
 94  0.96387D+00
 95  -.63862D+00
 96  0.20067D+00
 97  0.27268D+00
 98  -.69792D+00
 99  0.10000D+01
100  0.00000D+00
101  0.46305D+00
102  -.84438D+00
103  0.10767D+01
104  -.11190D+01
105  0.96387D+00
106  -.63862D+00
107  0.20067D+00
108  0.27269D+00
109  -.69793D+00
110  0.10000D+01
111  0.00000D+00
112  0.46305D+00
113  -.84438D+00
114  0.10767D+01
115  -.11190D+01
116  0.96387D+00
117  -.63861D+00
118  0.20067D+00
119  0.27269D+00
120  -.69793D+00
121  0.10000D+01
```

6.4.3 *CD equation*

Results corresponding to the fourth-order RFB bubble function and $D_a = 50$. Mesh scheme BLNR1 and boundary conditions BC2.

```
TITLE : FIELD VARIABLE

RESULT (NODE NO. , FIELD VARIABLE)

     1 0.00000D+00
     2 0.50738D-12
     3 0.12187D-10
     4 0.28315D-09
     5 0.65425D-08
     6 0.15130D-06
     7 0.34984D-05
     8 0.80892D-04
     9 0.18704D-02
    10 0.43248D-01
    11 0.10000D+01
    12 0.00000D+00
    13 0.50973D-12
    14 0.12216D-10
    15 0.28292D-09
    16 0.65438D-08
    17 0.15130D-06
    18 0.34984D-05
    19 0.80892D-04
    20 0.18704D-02
    21 0.43248D-01
    22 0.10000D+01
    23 0.00000D+00
    24 0.48366D-12
    25 0.12298D-10
    26 0.28267D-09
    27 0.65444D-08
    28 0.15130D-06
    29 0.34984D-05
    30 0.80892D-04
    31 0.18704D-02
    32 0.43248D-01
    33 0.10000D+01
    34 0.00000D+00
    35 0.52707D-12
```

CD equation, PE = 50, fourth-order RFB bubble function, mesh scheme BLNR1, and boundary conditions BC2 (see Fig. 4.8).

```
 36  0.12121D-10
 37  0.28337D-09
 38  0.65418D-08
 39  0.15131D-06
 40  0.34984D-05
 41  0.80892D-04
 42  0.18704D-02
 43  0.43248D-01
 44  0.10000D+01
 45  0.00000D+00
 46  0.52329D-12
 47  0.12191D-10
 48  0.28295D-09
 49  0.65439D-08
 50  0.15130D-06
 51  0.34984D-05
 52  0.80892D-04
 53  0.18704D-02
 54  0.43248D-01
 55  0.10000D+01
 56  0.00000D+00
 57  0.44019D-12
 58  0.12421D-10
 59  0.28233D-09
 60  0.65455D-08
 61  0.15129D-06
 62  0.34984D-05
 63  0.80891D-04
 64  0.18704D-02
 65  0.43248D-01
 66  0.10000D+01
 67  0.00000D+00
 68  0.58771D-12
 69  0.11938D-10
 70  0.28392D-09
 71  0.65401D-08
 72  0.15131D-06
 73  0.34984D-05
 74  0.80892D-04
 75  0.18704D-02
 76  0.43248D-01
 77  0.10000D+01
 78  0.00000D+00
 79  0.46776D-12
 80  0.12361D-10
 81  0.28247D-09
 82  0.65451D-08
 83  0.15129D-06
 84  0.34984D-05
 85  0.80892D-04
```

```
 86  0.18704D-02
 87  0.43248D-01
 88  0.10000D+01
 89  0.00000D+00
 90  0.47241D-12
 91  0.12315D-10
 92  0.28265D-09
 93  0.65446D-08
 94  0.15130D-06
 95  0.34984D-05
 96  0.80892D-04
 97  0.18704D-02
 98  0.43248D-01
 99  0.10000D+01
100  0.00000D+00
101  0.58953D-12
102  0.11955D-10
103  0.28378D-09
104  0.65409D-08
105  0.15131D-06
106  0.34984D-05
107  0.80892D-04
108  0.18704D-02
109  0.43248D-01
110  0.10000D+01
111  0.00000D+00
112  0.42702D-12
113  0.12477D-10
114  0.28220D-09
115  0.65456D-08
116  0.15129D-06
117  0.34984D-05
118  0.80892D-04
119  0.18704D-02
120  0.43248D-01
121  0.10000D+01
```

6.4.4 *CDR equation*

Results corresponding to the second-order STC bubble function and $D_a = -10$ and $P_e = 10$. Mesh scheme BLNR1 and boundary conditions BC2.

```
TITLE : FIELD VARIABLE

RESULT (NODE NO., FIELD VARIABLE)

  1  0.00000D+00
  2  0.42453D-08
  3  0.38208D-07
  4  0.32423D-06
```

```
5 0.27414D-05
6 0.23173D-04
7 0.19588D-03
8 0.16557D-02
9 0.13996D-01
10 0.11830D+00
11 0.10000D+01
12 0.00000D+00
13 0.42453D-08
14 0.38208D-07
15 0.32423D-06
16 0.27414D-05
17 0.23173D-04
18 0.19588D-03
19 0.16557D-02
20 0.13996D-01
21 0.11830D+00
22 0.10000D+01
23 0.00000D+00
24 0.42453D-08
25 0.38208D-07
26 0.32423D-06
27 0.27414D-05
28 0.23173D-04
29 0.19588D-03
30 0.16557D-02
31 0.13996D-01
32 0.11830D+00
33 0.10000D+01
34 0.00000D+00
35 0.42453D-08
36 0.38208D-07
37 0.32423D-06
38 0.27414D-05
39 0.23173D-04
40 0.19588D-03
41 0.16557D-02
42 0.13996D-01
43 0.11830D+00
```

CDR equation, $D_a = -10$, $P_e = 10$, exponential regime, second-order bubble function. Mesh scheme BLNR1 and boundary conditions BC2.

```
44  0.10000D+01
45  0.00000D+00
46  0.42453D-08
47  0.38208D-07
48  0.32423D-06
49  0.27414D-05
50  0.23173D-04
51  0.19588D-03
52  0.16557D-02
53  0.13996D-01
54  0.11830D+00
55  0.10000D+01
56  0.00000D+00
57  0.42453D-08
58  0.38208D-07
59  0.32423D-06
60  0.27414D-05
61  0.23173D-04
62  0.19588D-03
63  0.16557D-02
64  0.13996D-01
65  0.11830D+00
66  0.10000D+01
67  0.00000D+00
68  0.42453D-08
69  0.38208D-07
70  0.32423D-06
71  0.27414D-05
72  0.23173D-04
73  0.19588D-03
74  0.16557D-02
75  0.13996D-01
76  0.11830D+00
77  0.10000D+01
78  0.00000D+00
79  0.42453D-08
80  0.38208D-07
81  0.32423D-06
82  0.27414D-05
83  0.23173D-04
84  0.19588D-03
85  0.16557D-02
86  0.13996D-01
87  0.11830D+00
88  0.10000D+01
89  0.00000D+00
90  0.42453D-08
91  0.38208D-07
92  0.32423D-06
93  0.27414D-05
```

```
 94  0.23173D-04
 95  0.19588D-03
 96  0.16557D-02
 97  0.13996D-01
 98  0.11830D+00
 99  0.10000D+01
100  0.00000D+00
101  0.42453D-08
102  0.38208D-07
103  0.32423D-06
104  0.27414D-05
105  0.23173D-04
106  0.19588D-03
107  0.16557D-02
108  0.13996D-01
109  0.11830D+00
110  0.10000D+01
111  0.00000D+00
112  0.42453D-08
113  0.38208D-07
114  0.32423D-06
115  0.27414D-05
116  0.23173D-04
117  0.19588D-03
118  0.16557D-02
119  0.13996D-01
120  0.11830D+00
121  0.10000D+01
```

Appendices

A.1 Cartesian and cylindrical coordinate systems

The position vectors for points a_1 and a_2 in a Cartesian coordinate system shown in Fig. A.1 are given as

$$\mathbf{d}_1 = x_1\boldsymbol{i} + y_1\boldsymbol{j} + z_1\boldsymbol{k}, \quad \mathbf{d}_2 = x_2\boldsymbol{i} + y_2\boldsymbol{j} + z_2\boldsymbol{k},$$

where i, j, and k are unit vectors in the x, y, and z directions, respectively. The magnitude of line a_1a_2 (i.e. the distance between points $a_1$and a_2) can also be defined as

$$\overline{a_1a_2} = \sqrt{(x_2 - x_1)^2 + (y_2 - y_1)^2 + (z_2 - z_1)^2}.$$

For the vector $\mathbf{d}= xi + yi + zk$ the direction cosines are the numbers and are defined as

$$\begin{cases} \cos\alpha = \dfrac{x}{|d|} \\ \cos\beta = \dfrac{y}{|d|} \qquad [\cos^2\alpha + \cos^2\beta + \cos^2\gamma = 1], \\ \cos\gamma = \dfrac{z}{|d|} \end{cases}$$

where α, β, and γ are the angles which the vector makes with the positive directions of the coordinate axis shown in Fig. A.1 and $|\mathbf{d}| = \sqrt{x^2 + y^2 + z^2}$ is the vector magnitude. Another commonly used orthogonal coordinate system is the cylindrical system (Fig A.2)

Cylindrical (r, θ, z)

$$\begin{cases} x = d\cos\theta \\ y = d\sin\theta \\ z = z \end{cases} \quad \text{or} \quad \begin{aligned} d &= \sqrt{x^2 + y^2} \\ \theta &= \tan^{-1}(y/x) \\ z &= z \end{aligned}$$

In the system of o 123, under rotation of the coordinate system to $o\ \ \bar{1}\bar{2}\bar{3}$ orthogonal transformation of a vector $\boldsymbol{A}$ with components A_1, A_2, and A_3

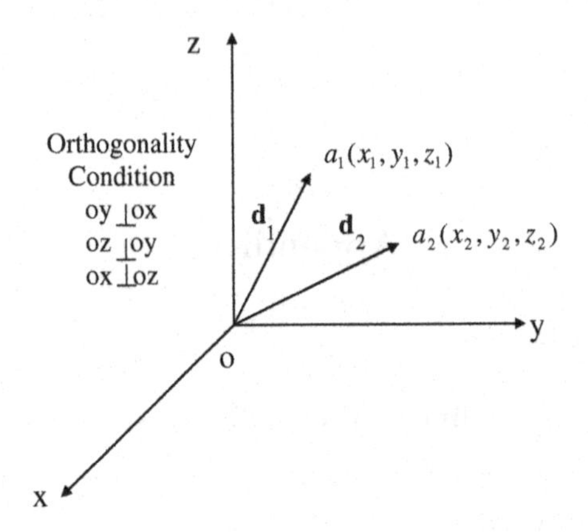

Fig. A.1.

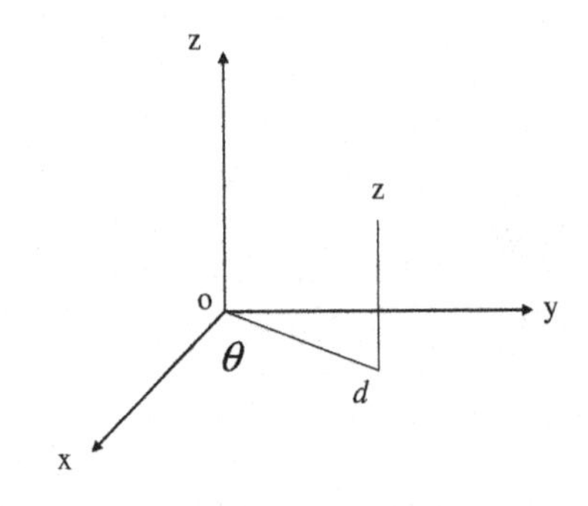

Fig. A.2.

is expressed by the following equation in the Cartesian system:

$\overline{A_J} = \mathrm{d}_{JI} A_I$(summation convention over the repeated index I is implied),

in which $\overline{A_J}$ are the components of the vector in coordinate system $o\ \bar{1}\bar{2}\bar{3}$ and d_{JI} identifies the cosine of the angles between the old axis oi ($i = 1, 2, 3$) and the new one $o\bar{j}$. Therefore,

$$\begin{Bmatrix} \overline{A_1} \\ \overline{A_2} \\ \overline{A_3} \end{Bmatrix} = \begin{Bmatrix} d_{11} & d_{12} & d_{13} \\ d_{21} & d_{22} & d_{23} \\ d_{31} & d_{32} & d_{33} \end{Bmatrix} \begin{Bmatrix} A_1 \\ A_2 \\ A_3 \end{Bmatrix}.$$

If a vector remains unchanged in such a transformation (i.e. $\overline{\mathbf{A}} = \boldsymbol{A}$), it is said to be invariant.

A.2 Some vector calculus relations

Let: $\mathbf{A} = \mathbf{A}\ (x, y, z)$ then $d\mathbf{A} = \frac{\partial \mathbf{A}}{\partial x}dx + \frac{\partial \mathbf{A}}{\partial y}dy + \frac{\partial \mathbf{A}}{\partial z}dz$.

For a vector $\mathbf{Q}$ where

$$\mathbf{Q}(t) = x(t)I + y(t)j + z(t)k,$$

in which t is a scalar variable, we have

$$\frac{d\mathbf{Q}}{dt} = \frac{dx}{dt}i + \frac{dy}{dt}j + \frac{dz}{dt}k.$$

For differentiable vector functions **A, B**, and **C** and differentiable function F we have

$$\frac{d}{dt}(\mathbf{A} + \mathbf{B}) = \frac{d\mathbf{A}}{dt} + \frac{d\mathbf{B}}{dt},$$

$$\frac{d}{dt}(F\mathbf{A}) = \mathrm{F}\frac{d\mathbf{A}}{dt} + \frac{d\mathrm{F}}{dt}\mathbf{A},$$

$$\frac{d}{dt}(\mathbf{A} \cdot \mathbf{B}) = \mathbf{A} \cdot \frac{d\mathbf{B}}{dt} + \frac{d\mathbf{A}}{dt} \cdot \mathbf{B},$$

$$\frac{\partial}{\partial t}(\mathbf{A} \cdot \mathbf{B}) = \mathbf{A} \cdot \frac{\partial \mathbf{B}}{\partial t} + \frac{\partial \mathbf{A}}{\partial t} \cdot \mathbf{B},$$

$$\frac{d}{dt}(\mathbf{A} \times \mathbf{B}) = \mathbf{A} \times \frac{d\mathbf{B}}{dt} + \frac{d\mathbf{A}}{dt} \times \mathbf{B}, \quad \text{and}$$

$$\frac{\partial}{\partial t}(\mathbf{A} \times \mathbf{B}) = \mathbf{A} \times \frac{\partial \mathbf{B}}{\partial t} + \frac{\partial \mathbf{A}}{\partial t} \times \mathbf{B}.$$

The vector differential operator *del* (or *nabla*) written as ∇ is defined by

$$\nabla \equiv \quad \frac{\partial}{\partial x}i + \frac{\partial}{\partial y}j + \frac{\partial}{\partial z}k \equiv i\frac{\partial}{\partial x} + j\frac{\partial}{\partial y} + k\frac{\partial}{\partial z},$$

where i, j, and k are the unit vectors in a Cartesian coordinate system.

The *gradient* of a scalar $F\ (x, y, z)$ is defined by

$$\nabla(F) = \frac{\partial F}{\partial x}i + \frac{\partial F}{\partial y}j + \frac{\partial F}{\partial z}k.$$

The *divergence* of a vector $\boldsymbol{A}(x, y, z)$ is defined by

$$\nabla \cdot \boldsymbol{A} = \left(\frac{\partial}{\partial x}i + \frac{\partial}{\partial y}j + \frac{\partial}{\partial z}k\right) \cdot (A_1 i + A_2 j + A_3 k) = \frac{\partial A_1}{\partial x} + \frac{\partial A_2}{\partial y} + \frac{\partial A_3}{\partial z}$$

(note that $\nabla \cdot \boldsymbol{A} \neq \boldsymbol{A} \cdot \nabla$).

The *curl* of a vector $\boldsymbol{A}(x, y, z)$ is defined by

$$\nabla \times \mathbf{A} = \left(\frac{\partial}{\partial x}i + \frac{\partial}{\partial y}j + \frac{\partial}{\partial x}k\right) \times (A_1 i + A_2 j + A_3 k)$$
$$= \begin{vmatrix} i & j & k \\ \dfrac{\partial}{\partial x} & \dfrac{\partial}{\partial y} & \dfrac{\partial}{\partial z} \\ A_1 & A_2 & A_3 \end{vmatrix},$$
$$\nabla \times (\nabla \times \mathbf{A}) = \nabla(\nabla \cdot \mathbf{A}) - \nabla^2 \mathbf{A},$$

where

$$\nabla^2 = \frac{\partial^2}{\partial x^2} + \frac{\partial^2}{\partial y^2} + \frac{\partial^2}{\partial z^2}$$

and (in a Cartesian system) is called the *Laplacian* operator.

A.3 Definition of a vector space

A vector space consists of four parts:

- A set of vectors $\boldsymbol{V}$
- A set of scalars F (either all real numbers R or all complex numbers **C**)
- A rule,+, for adding vectors in $\boldsymbol{V}$
- A rule, ·, for multiplying vectors in $\boldsymbol{V}$ by scalars on F.

If the above requirements are satisfied then ($\boldsymbol{V}$, F, +.) makes a vectors space if, and only if, the conditions below are satisfied.

Vector addition operation must satisfy the following axioms:

For all **u, v**, and **w** in $\boldsymbol{V}$:

$$\mathbf{u} + \mathbf{v} = \mathbf{v} + \mathbf{u},$$
$$(\mathbf{u} + \mathbf{v}) + \mathbf{w} = \mathbf{u} + (\mathbf{v} + \mathbf{w}).$$

A vector denoted by **0**, in $\boldsymbol{V}$ which satisfies

$$\mathbf{u} + \mathbf{0} = \mathbf{u}, \quad \text{for every } \mathbf{u} \text{ in } \boldsymbol{V}.$$

is called the zero vector in $\boldsymbol{V}$. For each vector **u** in $\boldsymbol{V}$, there is a vector, denoted by $-\boldsymbol{u}$, in $\boldsymbol{V}$ such that

$$\mathbf{u} + (-\mathbf{u}) = 0,$$

$-\mathbf{u}$ is called the additive inverse of **u**.

Multiplication by scalar: for all vectors **u**, **v** in $\boldsymbol{V}$ and all scalars r, s, t in F,

$$1\mathbf{u} = \mathbf{u}, \qquad (\mathrm{st})\mathbf{u} = \mathrm{s}(t\mathbf{u}),$$
$$\mathrm{r}(\mathbf{u}+\mathbf{v}) = \mathrm{r}\mathbf{u} + \mathrm{r}\mathbf{v}, \quad (\mathrm{s}+\mathrm{t})\mathbf{u} = \mathrm{s}\mathbf{u} + \mathrm{t}\mathbf{u}.$$

If F represents the set of all real numbers then ($\boldsymbol{V}$, R, +,.) is called a real vectors space, whereas if F is the set of all complex numbers, then ($\boldsymbol{V}$, C, +,.) is called a complex vector space.

A.3.1 Subspaces

Usually for a given problem the underlying vector space is known; for example, it may be R^n or C^n. In general only a subset of vectors from this vector space is used for the solution of a given problem. This subset of vectors is itself a vector space under the same operation of addition and scalar multiplication as in V if, and only if, it is closed under addition and closed under scalar multiplication.

A.3.2 Spanning sets

A set of vectors $\{\mathbf{v}_1, \mathbf{v}_2, \ldots, \mathbf{v}_k\}$ in a vector space spans V if every vector in $\boldsymbol{V}$ can be written as a linear combination of $\mathbf{v}_1, \mathbf{v}_2, \ldots, \mathbf{v}_k$, that is if for any $\mathbf{v} \in \mathbf{V}$ there exist scalars $a_1, a_2, \ldots, a_k$, such that

$$\mathbf{v} = a_1\mathbf{v}_1 + a_2\mathbf{v}_2 + \cdots a_k\mathbf{v}_k.$$

For any given set of vectors $\{\mathbf{v}_1, \mathbf{v}_2, \ldots, \mathbf{v}_k\}$ in a vector space $\boldsymbol{V}$, the set of all vectors that can be written as a linear combination of $\mathbf{v}_1, \mathbf{v}_2, \ldots, \mathbf{v}_k$, can be formed. This set of vectors is a subspace of $\boldsymbol{V}$ called the *subspace spanned by* $\{\mathbf{v}_1, \mathbf{v}_2, \ldots, \mathbf{v}_k\}$ and indicated by span $\{\mathbf{v}_1, \mathbf{v}_2, \ldots, \mathbf{v}_k\}$. Therefore, span $\{\mathbf{v}_1, \mathbf{v}_2, \ldots, \mathbf{v}_k\} = \{\mathbf{v} \in \mathbf{V} : \mathbf{V} = a_1\mathbf{v}_1 + a_2\mathbf{v}_2 + \cdots + a_k\mathbf{v}_k\}$.

A.4 Divergence (Gauss) theorem

For a volume V bounded by a closed surface S, if $\boldsymbol{A}$ is a vector function of position with continuous derivatives, then

$$\iiint_V \nabla \cdot \mathbf{A} dV = \iint_S \mathbf{A} \cdot n dS = \oint_S \int \mathbf{A} \cdot dS.$$

The surface integral $\iint_S \mathbf{A} \cdot dS$ denotes the flux of $\boldsymbol{A}$ over the closed surface S.

A.5 Stokes theorem

For an open two-sided surface S bonded by a curve C, the line integral of vector $\boldsymbol{A}$ (curve C is traversed in the positive direction) is expressed as

$$\oint_C \mathbf{A} \cdot dr = \iint_S (\nabla \times \mathbf{A}) \cdot n dS = \iint_S (\nabla \times \mathbf{A}) \cdot dS.$$

A special case of the Stokes theorem expressed as

$$\oint_L [f_1(x,y)dx + f_2(x,y)dy] = \iint_R \left(\frac{\partial f_2}{\partial x} - \frac{\partial f_1}{\partial y} \right) dxdy$$

is called Green's theorem (in a plane). The Gauss divergence theorem can be viewed as the generalization of Green's theorem and is obtained by replacing planar region R and its boundary curve C with a three-dimensional region and its closing surface.

A.6 Green's function

A Green's function, $G(x, s)$, of a linear differential operator $L = L(x)$ over a subset of the Euclidean space R^n, is any solution of

$$LG(x,s) = \delta(x - s),$$

where δ is the Dirac delta function. This technique can be used to solve differential equations of the form

$$LT(x) = f(x).$$

If a function G can be found for the operator L, then after the multiplication of the differential equation by $f(s)$ and its integration with respect to s such that

$$\int LG(x,s)f(s)ds = \int \delta(x - s)f(s)ds = f(x).$$

Then $f(x)$ in the right-hand side can be replaced to obtain

$$LT(x) = \int LG(x,s)f(s)ds.$$

Because the operator $L = L(x)$ is linear and is acting on variable x alone (not on the variable of integration s), we can write

$$LT(x) = L\left(\int G(x,s)f(s)ds \right)$$

or

$$T(x) = \int G(x,s)f(s)ds.$$

A.7 Smooth functions

A smooth function is a function that has continuous derivatives up to some order over an interval $[a, b]$. The number of continuous derivatives necessary for a function to be considered smooth is problem dependent and may vary from two to infinity. A function for which all orders of derivatives are continuous is called a C^∞ function. For example,

$$f(x) = e^{2x}$$

is a C^∞ function because its nth derivative expressed as

$$f^{(n)}(x) = 2^n e^{2x},$$

exists and is continuous.

A.8 Self-adjoint and symmetric operators

Consider a second-order differential operator:

$$LT(x) = A_0 \frac{d^2T}{dx^2} + A_1 \frac{dT}{dx} + A_2 T,$$

where T and A_i are real functions of x on the region of interest $[a, b]$ with continuous derivatives and $A_0(x) \neq 0$ on $[a, b]$. This means that there are no singular points in $[a, b]$. Then an *adjoint* operator L^* is defined by

$$L^* = \frac{d^2}{dx^2}(A_0 T) - \frac{d}{dx}(A_1 T) + A_2 T,$$

$$L^* = A_0 \frac{d^2T}{dx^2} + (2A_0' - A_1)\frac{dT}{dx} + (A_0'' - A_1' + A_2)T.$$

In a *self-adjoint* operator, the operator is equal to its adjoint

$$L = L^*.$$

It also means that

$$A_0'(x) = A_1(x) \quad \text{and} \quad A_0'''(x) = A_1'(x).$$

A linear operator L on a Hilbert space H with dense domain D_L called symmetric if and only if for $f, g \in D_L$

$$(Lf, g) = (Lg, f).$$

If $D_L = H$ then L is a bounded operator and is self-adjoint.

Author Index

Subject Index